Case Studies on Food Experiences in Marketing, Retail, and Events

Woodhead Publishing Series in Consumer Science and Strategic Marketing

Case Studies on Food Experiences in Marketing, Retail, and Events

Edited by

Susanne Doppler

Adrienne Steffen

Series Editors

Alessio Cavicchi

Cristina Santini

ELSEVIER

WP WOODHEAD PUBLISHING An imprint of Elsevier

Woodhead Publishing is an imprint of Elsevier
The Officers' Mess Business Centre, Royston Road, Duxford, CB22 4QH, United Kingdom
50 Hampshire Street, 5th Floor, Cambridge, MA 02139, United States
The Boulevard, Langford Lane, Kidlington, OX5 1GB, United Kingdom

Notices
Knowledge and best practice in this field are constantly changing. As new research and experience broaden our understanding, changes in research methods, professional practices, or medical treatment may become necessary.

Practitioners and researchers must always rely on their own experience and knowledge in evaluating and using any information, methods, compounds, or experiments described herein. In using such information or methods they should be mindful of their own safety and the safety of others, including parties for whom they have a professional responsibility.

To the fullest extent of the law, neither the Publisher nor the authors, contributors, or editors, assume any liability for any injury and/or damage to persons or property as a matter of products liability, negligence or otherwise, or from any use or operation of any methods, products, instructions, or ideas contained in the material herein.

British Library Cataloguing-in-Publication Data
A catalogue record for this book is available from the British Library

Library of Congress Cataloging-in-Publication Data
A catalog record for this book is available from the Library of Congress

ISBN: 978-0-12-817792-1 (print)

ISBN: 978-0-12-817793-8 (online)

For information on all Woodhead Publishing publications
visit our website at https://www.elsevier.com/books-and-journals

Publisher: Charlotte Cockle
Acquisitions Editor: Megan R. Ball
Editorial Project Manager: Lena Sparks
Production Project Manager: Vijayaraj Purushothaman
Cover Designer: Alan Studholme

Typeset by MPS Limited, Chennai, India

Contents

List of contributors

Dilip Dinkar Ambarkhane Symbiosis School of Banking and Finance, Symbiosis International (Deemed University), Pune, India

Claudia Bauer-Krösbacher International Business, IMC University of Applied Sciences Krems, Krems, Austria

Dajana Baumgartner International Business, IMC University of Applied Sciences Krems, Krems, Austria

Giovanna Bertella School of Business and Economics, UiT The Arctic University of Norway, Tromsø, Norway

Peter Bjork Department of Marketing, CERS, Hanken School of Economics, Vaasa, Finland

Antonella Capriello Department of Economics and Business Studies, University of Piemonte Orientale (UPO), Novara, Italy

Mauricio Cepeda Department of Marketing, Business School, Universidad San Francisco de Quito, Quito, Ecuador; Department of Hospitality Management, School of Hospitality Management and Culinary Arts, Universidad San Francisco de Quito, Quito, Ecuador

Susanne Doppler Event Management and Tourism, Hochschule Fresenius Heidelberg, Heidelberg, Germany; University of Applied Sciences Fresenius Heidelberg, Heidelberg, Germany

Dennis Hanson The Federal University of Mato Grosso do Sul (UFMS), School of Engineering—FAENG, Pioneiros, Brazil

Rohail Hassan Othman Yeop Abdullah Graduate School of Business (OYAGSB), Universiti Utara Malaysia (UUM), Sintok, Malaysia

M. Muzakkeerul Huda Presidency University, Dhaka, Bangladesh

Tina Umnia Hussain Presidency University, Dhaka, Bangladesh

Hannele Kauppinen-Räisänen School of Marketing and Communication, University of Vaasa, Vaasa, Finland

Eirini Koronaki Department of Business Administration of Food and Agricultural Enterprises, University of Patras, Patras, Greece

Prabhat Kumar Bharati Vidyapeeth (Deemed University), Pune, India

Attila Pohlmann Department of Marketing, Business School, Universidad San Francisco de Quito, Ecuador

Abu Saleh Cranberra Business School, University of Canberra, Australia

Ardhendu Shekhar Singh Symbiosis School of Banking and Finance, Symbiosis International (Deemed University), Pune, India

Adrienne Steffen Consumer Behavior and Market Research, Hochschule Fresenius Heidelberg, Heidelberg, Germany; University of Applied Sciences Fresenius Heidelberg, Heidelberg, Germany

Tony Stovall Kelley School of Business, Indiana University, Indianapolis, IN, United States

Parves Sultan School of Business and Law, Central Queensland University, Melbourne, VIC, Australia

Tasmiha Tarafder RMIT University, Melbourne, Australia

Prokopis K. Theodoridis Department of Business Administration of Food and Agricultural Enterprises, University of Patras, Patras, Greece

Stephanie Tischler International Business, IMC University of Applied Sciences Krems, Krems, Austria

Mine Üçok-Hughes College of Business and Economics, California State University, Los Angeles, Los Angeles, CA, United States

Franklin Velasco-Vizcaíno Department of Marketing, Business School, Universidad San Francisco de Quito, Quito, Ecuador

Rossen Ventzislavov College of Liberal Arts, Woodbury University, Burbank, CA, United States

Ho Yin Wong Department of Marketing, Deakin University, Burwood, VIC, Australia

Lisa-Marie Wurzer University of Applied Sciences Fresenius Heidelberg, Heidelberg, Germany

Biography

Dr. Dilip Ambarkhane, PhD (Bharati Vidyapeeth, Pune) and alumni of IIT (Bombay), is working as an associate professor at Symbiosis School of Banking and Finance, India. He also worked as visiting faculty at University of Pune and National Institute of Bank Management, Pune, India. He has 36 years' experience of working in a Public Sector Bank. He has published papers in *International Journal of Rural Management*, *Indian Growth and Development Review*, etc. and has presented as well as acted as track chair in many international conferences. He is a coeditor for a special issue on Financial Inclusion for *International Journal of Business and Globalization*.

Claudia Bauer-Kröbacher is a programme director of the Bachelor programme "Tourism and Leisure Management" at the IMC University of Applied Sciences Krems in Austria. She is also involved in lecturing and research. She received her PhD from the Dublin Institute of Technology, holds an MA in Economics from the University of Innsbruck, and Prof. (FH) from the IMC University of Applied Sciences Krems. Her research interests include cultural tourism, authenticity, tourism consumer behavior, multivariate, and qualitative research methods. Her research has been published in the *Journal of Travel Research* and several books. Claudia also has extensive experience as a lecturer and in didactics.

Dr. Peter Björk is a professor of marketing at HANKEN School of Economics, Vaasa, Finland. His research focus is in the field of service marketing, consumer behavior, tourism marketing, and destination development. He has a special interest in service innovation and design, destination branding, tourist experiences, and attraction marketing. Björk is an associate editor for *Scandinavian Journal of Hospitality and Tourism* and *Finnish Journal of Tourism Research*. He has authored articles published in *Tourism Management*, *International Journal of Contemporary Hospitality Management*, *Tourism Management Perspectives*, and *Journal of Business Research*, and he is an ad hoc reviewer form more than 10 marketing-related journals.

Antonella Capriello, PhD, is an associate professor of Marketing at the University of Eastern Piedmont in Italy. The research activities include studies on event management, networking processes in tourist destinations, social entrepreneurship, and franchising. She published more than 60 research papers, including articles in the *Journal of Business Research* and *Tourism Management*.

Dr. Susanne Doppler is a professor of Event Management and Tourism at the Hochschule Fresenius Heidelberg and its predecessor institution since 2012. She studied at the University of Hohenheim in Germany where she researched ecological aspects of sustainability in cooperation with the Universities of Hohenheim and Stuttgart. She has published in the areas of experience quality, sustainable business practice-s, and stress in event management. Further areas of interests are future of events, food events and destination marketing, and the transformation economy.

Prabhat Kumar, an alumnus of the Indian Institute of Management Ahmadabad and University of Pune, India, has been an associate professor of Management at Bharati Vidyapeeth University, Pune for 10 years. He worked with IT major Hewlett Packard (previously Electronic Data Systems) as an HR Unit Manager and with staffing behemoth Manpower (previously ABC Consultants) as a consultant before moving to academics in 2010. He has expertise in Human Resource Management and his area of research include Social Responsibilities of Small and Medium-sized Enterprises and their sustainability, Alternative Food Networks, and Inclusive Higher Education.

Dennis Hanson is an adjunct professor at the Faculty of Engineering at UFMS, PhD in Business at Anhaguera Uniderp University (2019) with research in Design Thinking Applied to the Hotel Business in Bonito, Mato Grosso do Sul, Brazil, a master in Production Engineering with research in marketing at Universidade Federal Fluminense (2004), and graduated in Administration at Universidade Paulista São José. He has technical training in the visual arts area at the School of Visual Arts in Rio de Janeiro and long experience in the area of Design and Business Management. His areas of interest are business strategy, strategic management, business management, innovation, marketing, design, design thinking, and strategic communication management.

Dr. Rohail Hassan is a senior lecturer of Corporate Governance and Finance at Othman Yeop Abdullah Graduate School of Business, Universiti Utara Malaysia. He is acting in the capacity of Series Editor titled *"Big Data for Industry 4.0: Challenges and Applications—Taylor & Francis Group.* His research is in corporate governance, diversity and Inclusion, big data, marketing analytics, firm performance, and strategy. Rohail has published research papers in leading management journals such as *Journal of Management and Organization, Journal of Cleaner Production, Journal of Intellectual Capital, and Economics Bulletin.* In total, he has published over 40 refereed journal articles to date.

M. Muzakkeerul Huda is an academic staff in the School of Business, Presidency University, Dhaka, Bangladesh. He has published research papers in *International Review of Business Research Papers*, *Indian Journal of Marketing*, *International Journal of Social Business*, *International Journal of Marketing Financial Services and Management Research*, *Journal of Business and Economics*, *Journal of Business and Technology*, and *Journal of Business Administration*.

Tina Umnia Hussain is an academic staff in the School of Business, Presidency University, Dhaka, Bangladesh. She has published a paper in an International Conference in Dhaka and had also coauthored a book chapter on entrepreneurship. She can be contacted at: tina.umnia@gmail.com.

Dr. Hannele Kauppinen-Räisänen works as an associate professor at the University of Vaasa, Finland. She has held positions at Hanken School of Economics, Finland and the International University of Monaco. Her research covers issues such as tourism and hospitality, products, and service design. She has published in academic journals such as *Tourism Management*, *International Journal of Contemporary Hospitality Management*, *Current issues in Tourism*, *Scandinavian Journal of Tourism and Hospitality*, *Journal of Service Management*, and *Journal of Business Research*. She has received several awards for her research. She is an active scholar who publishes widely and works closely with firms on various marketing issues.

Eirini Koronaki is a postdoctoral researcher at the Department of Business Administration of Food and Agricultural Enterprises at the University of Patras on the subject of sustainability in luxury consumption. Her research interests lie in the fields of Luxury Brand Management, Consumer Behavior, International Marketing, and Marketing Communications. She has coauthored book chapters and has presented papers in several conferences, while she has published articles in peer-reviewed academic journals. She has teaching experience on an undergraduate, postgraduate, and executive level. She has taken part in consultancy programs for business executives and has professional experience as a Business Development Manager at Eurodiet Med.

Attila Pohlmann is a professor of marketing at the Business School at Universidad San Francisco de Quito in Ecuador. His research focuses on transformative consumer research, specifically investigating the roles of marketing and consumption in creating, and possibly solving, societal, and environmental problems. He published in the *Journal of Interactive Marketing*, *Industrial Marketing Management*, *Critical Studies in Men's Fashion*, and *Journal of Consumer Marketing*. Find out more about community outreach, teaching, and other projects at http://www.attila-pohlmann.com.

Dr. Abu Saleh is an associate professor at the University of Canberra, Australia. He published research papers in *Management Decision*, *Journal of Business Research*, *Australasian Marketing Journal*, *International Journal of Bank Marketing*, and *Asian Social Science*.

Ardhendu Shekhar Singh, Fellow (IRMA), is an associate professor at Symbiosis School of Banking and Finance, Symbiosis International (Deemed University). He has published papers in international journals such as *Qualitative Market Research: an International Journal*, *Indian Growth and Development Review*, etc. and has presented as well as acted as track chair in many international conferences. He is Editorial Board Member at *International Journal of Marketing and Sales Education* and *Journal of Global Business Insights*. He is editing a special issue on Financial Inclusion for *International Journal of Business and Globalization*.

Dr. Adrienne Steffen is a professor of Consumer Behaviour and Market Research at the Hochschule Fresenius Heidelberg and its predecessor institution since September 2012. She received her BBA in International Management at the International University in Germany and studied at the University of Michigan in the United States and the ESC Rennes in France. She completed her PhD and postgraduate diploma in Research in Business Management at the University of Strathclyde in Glasgow. She has recently published in the area of consumer marketing where she focuses on experience management, food marketing, and sustainable consumption.

Tony Stovall's research is broadly centered around issues of consumer well-being, consumption experiences, and digital consumption habits. He has been an assistant professor of Marketing at Woodbury University's School of Business in Burbank, California, United States, and Towson University in Towson, Maryland, United States. He is currently on the faculty of The Kelley School of Business at Indiana University, Indianapolis, United States. Some of the outlets in which he has published include the *Journal of Public Policy and Marketing; Consumption, Markets and Culture*; *Electronic Markets*; *Journal of Marketing Management*; and *N.A. Advances in Consumer Research.*

Dr. Parves Sultan is a senior lecturer in Marketing at the Central Queensland University, Australia. His research articles appeared in several scholarly journals, including the *Journal of Brand Management*, *Journal of Business and Industrial Marketing*, *Managing Service Quality (Journal of Service Theory and Practice)*, *International Journal on Disability and Human Development*, *Asia Pacific Journal of Marketing and Logistics*, *Australasian Journal of Regional Studies*, *Quality Assurance in Education*, and *International Journal of Quality and Service Sciences* are worth mentioning.

Tasmiha Tarafder is a PhD candidate at the RMIT University, Australia. She had obtained her master by research from the University of Canberra, Australia. She received the Best paper award 2013 and the Highly Commended Award 2012 from the ANZSRAI conferences. She also published in the Proceedings of the 21st Biennial ASAA Conference, Australian National University, Canberra. She published a research paper in the *Australasian Journal of Regional Studies* and a book. She can be contacted at: tasmiha.tarafder@rmit.edu.au.

Prokopis K. Theodoridis is an associate professor of Marketing at the Department of Business Administration of Food and Agricultural Enterprises of the University of Patras. He holds an MSc in Marketing (University of Stirling, UK) and was awarded a PhD from the Athens University of Economics and Business. His primary research interests include the areas of consumer behavior, food and beverage marketing, marketing communications, retail and services marketing, and internal marketing. He is General Secretary of the Greek Marketing Academy and member of European Marketing Academy.

Stephanie Tischler is a PhD student and lecturer at the IMC University of Applied Sciences Krems, Austria. She teaches in the Bachelor programmes "Tourism and Leisure Management" and "International Wine Business." Stephanie graduated from the Vienna University of Economics and Business in business administration with specializations in tourism and leisure management and marketing science and business psychology. Her research focus is on quantitative and qualitative research methods, tourism marketing, consumer behavior, as well as health and culinary tourism.

Mine Üçok Hughes, PhD is an assistant professor of Marketing at California State University, Los Angeles, United States. She studies nonmainstream consumers, sustainable marketing, and social marketing. She presents regularly at marketing conferences and her articles have been published in various academic journals such as *Journal of Product and Brand Management*, *Business Horizons*, *Qualitative Market Research*, and *Advances in Consumer Research*.

Franklin Velasco has over 15 years of experience as a scholar. Additionally, Franklin has several years of experience in industry at companies such as Kraft, IBM, and IMAX. He obtained his PhD in Marketing at University of Texas at Arlington. He has received several awards such as 2019 Best Paper Award at INEKA Conference, 2017 Distinguished Doctoral Student, and 2017–2018 Research Award. Franklin research interests are in consumers' well-being, food marketing, sports marketing, and consumer psychology. His research has been published in high-profile journals, and he is a member of the Editorial Board for the *Journal of Business Research*.

Rossen Ventzislavov is a philosopher and cultural critic specializing in esthetics, architectural theory, literature, popular music, and performance art. He is the 2017 recipient of the *Curriculum Diversification Grant* from the American Society for Aesthetics. His work has appeared in *The Journal of Aesthetics and Art Criticism*, *Deleuze Studies*, *Contemporary Aesthetics*, *The Journal of Popular Music Studies*, etc. He is a professor of Philosophy at Woodbury University.

Dr. Ho Yin Wong is a senior lecturer at the Department of Marketing, Deakin University, Burwood Campus, Australia. He has published in the *International Marketing Review*, *Journal of Business and Industrial Marketing*, *Journal of Strategic Marketing*, *Journal of Global Marketing*, *Managing Service Quality*, *Marketing Intelligence and Planning*, *Journal of Brand Management*, *Journal of Product and Brand Management*, and *Asia Pacific Journal of Marketing and Logistics*. He can be contacted at: ho.wong@deakin.edu.au.

Lisa-Marie Wurzer graduated at the Hochschule Fresenius in Heidelberg, Germany. She studied International Business Management with a focus on Event-Management. In view of her bachelor thesis, she discussed and analyzed the topic of customer food experience using the example of Feinkost Käfer GmbH, Munich, Germany. She has a special interest in creating memorable holistic food experiences, which she now tries to implement in her daily business. She now works as a project assistant at Feinkost Käfer and is responsible for the conception and implementation of all kinds of events. Moreover she is part of the organization team of Käfer Wiesn Schänke at the Oktoberfest in Munich, Germany.

Introduction to the experience economy

1

Adrienne Steffen and Susanne Doppler
University of Applied Sciences Fresenius Heidelberg, Heidelberg, Germany

1.1 Introduction to the experience economy

The term "customer experience" has become a popular buzzword in the marketing literature (Lemon & Verhoef, 2016, p. 70) and has been described by marketers and academics as a strategic process (Jain, Aagja, & Bagdare, 2017, p. 642). The idea of creating a customer experience is not an entirely new concept because it relies on related research streams. Customer experience management (CEM) as a management discipline has recently received a lot of attention in theory and practice and has become an important marketing concept in the past three decades (Bruhn & Hadwich, 2012, p. 5).

The idea first arose in the mid-1980s (Doppler & Steffen, 2018; Steffen, 2012, p. 2). Hirschman and Holbrook (1982) wrote a paper on hedonic consumption which can be seen as a starting point of this movement (Bruhn & Hadwich, 2012, p. 5). The early concepts of customer experience all focused on the emotional customer gain. Consumers are motivated by the search for emotional arousal when selecting products to consume and seek *feelings and fun* (Hill & Gardner, 1986).

In the early 1990s when the concept gained popularity, it was a new management discipline (Carbone & Haeckel, 1994). At the end of the 1990s, Pine and Gilmore (1998) had a decisive influence on research with the insight that companies can gain a competitive advantage with their customer experience. A customer experience occurs "when a company intentionally uses services as the stage, and goods as props, to engage individual customers in a way that creates a memorable event" (Pine & Gilmore, 1998, p. 98). In their research they call the newly arising economy an experience economy.

One year later in 1999 Pine and Gilmore published a book titled *The Experience Economy* which proposes that the way people will consume in the future is changing. The typical differentiation advantages firms used up to then like price and product quality do not suffice anymore (p. 100). According to Pine and Gilmore (1998, p. 98), firms will require to not only create a customer experience but also stage an experience to create a competitive advantage. Product or service propositions are no longer adequate tools for differentiation advantages.

Shaw and Ivens (2002, p. 1) also took up on this idea and even called the customer experience a long-term competitive advantage. They further noted that evoking the right emotions might be one of the key success factors for companies to

Case Studies on Food Experiences in Marketing, Retail, and Events. DOI: https://doi.org/10.1016/B978-0-12-817792-1.00001-0

create customer experiences and 85% of senior business leaders agree that engaging consumers emotionally will increase customer loyalty more than basing the customer experience on physical characteristics (Shaw and Ivens, 2002, p. 62). Holbrook (2000) concluded that companies create customer experiences to educate, entertain, display knowledge, or offer an escape from reality. An analysis of photographic essays reveals that customer value is created by the three "F"s (fantasies, feelings, and fun) (Holbrook, 2006).

The customer experience should be as important as the product or service itself for the creation of a holistic customer experience (Carbone & Haeckel, 1994). Grönroos (2006) pointed out that the consumption experience itself is necessary to create customer value and not individual experience elements. Thus "all elements" such as information, encounters with other people, systems or infrastructure, and physical objects impact the customer's value creation process (Grönroos, 2006, p. 6). A good starting point for understanding the total customer experience is the assessment of the "clues" a firm is sending to its customers, for example, first concerning product functionality and second concerning the firms sensual environment such as smell and light (Berry, Carbone, & Haeckel, 2002, pp. 85–86).

To be able to create and deliver a unique total customer experience, companies require new competence profiles, for example, and to integrate several business functions such as information technology, logistics, marketing, service operations, human resources, and external partners (Lemon & Verhoef, 2016). This new competence profile is a long-term company goal which can be decided on after an "experience audit" has been undertaken (Berry et al., 2002, p. 86).

A further option of how firms get customers involved in the experience creation process is by letting them personalize or cocreate their products or service offerings. Thus the concept is clearly linked with value creation (Prahalad & Ramaswamy, 2004). Early examples were car manufacturers or sports companies such as Nike or Converse, which let customers cocreate and thus design their own cars or shoes. Today, companies in different industries allow their customers to be part of the customer experience by offering cocreation opportunities on their websites. Mars' popular M&Ms chocolate lenses can be customized with text and pictures for private occasions or used for business communication or gifts (Mymms, 2019). The popular Nivea Creme produced by Beiersdorf in the traditional blue tin with 75 mL can also be customized in the online shop with text and pictures (Nivea, 2018).

With increasing competitive intensity, the brand experience makes a decisive contribution to differentiation and profiling (Kirchgeorg & Ermer, 2012, p. 253), whereby a differentiation potential to be rated as high presupposes that the perceived experience quality and thus the emotional benefit of the customer experience is high (Jüttner, Schaffner, Windler, & Maklan, 2013). As Smit and Melissen (2018, p. 2) note in their book, "creating a brand experience, also through additional products and services and specific distribution choices, is now part of the game."

1.2 Definition and delineation of the customer experience from related marketing streams

1.2.1 Definition of the customer experience

Although marketers traditionally see humans as rational decision makers, marketers who believe in the experimental marketing approach see humans as both rational and emotional. They have a pleasurable experience when buying goods or using services (Hwang & Seo, 2016).

Batat (2019, p. 61) explains in his book that the customer experience is a "multi-dimensional, composite, and an evolving construct" which combines "a harmonious blend of several factors and actors that have a direct or indirect impact on designing successful experiences in connection with the marketplace or not" and has the consumer and the company perspective (p. 61). "Unlike the traditional marketing approach, which focuses on goods (products and services), experiential marketing offers a transformative experience where the consumer engages in a process of self-fulfillment" (p. 61).

By summarizing the term as used in the retail context described by, for example, Carbone and Haeckel (1994), Schmitt (1999), or Verhoef et al. (2009), customer experience can be defined as "the sum total of cognitive, emotional, sensorial, and behavioural responses produced during the entire buying process, involving an integrated series of interaction with people, objects, processes and environment in retailing" (Bagdare & Jain, 2013, p. 791).

The customer experience is generally portrayed as a short-lived personal, subjective experience from the customer side which is typically offered by the service provider. For this book, a holistic view on the customer experience is taken: "Customer experience is the sum of all impressions a customer has with a company over the span of his customer lifetime including individual perceptions, interactions and the quality of the company performance" (freely translated from Bruhn & Hadwich, 2012, p. 10).

The definition already shows that the customer experience has several components or dimensions. Bruhn and Hadwich (2012, p. 5) summarize in their review that research on the customer experience generally follows four research directions: the product experience, the service experience, the brand experience, and the consumption- or shopping experience.

It is suggested in the literature that no matter how ordinary the product or service encounter is, customers will have good, bad, or neutral experience (Carbone & Haeckel, 1994) and that all these experiences allow customers to engage emotionally (Berry & Carbone, 2007). All of these encounters (especially the positive ones) create the perception of value for customers and are determinants of brand or service preference (Berry & Carbone, 2007). The resulting challenge for firms is to engineer their customer experience (Carbone & Haeckel, 1994) so that customers engage positively, perceive value, and build brand preference.

This process is called CEM, which refers to "the process of strategically managing a customer's entire experience with a product or a company" (Schmitt, 2003, p. 17) along all customer touchpoints (Bruhn & Hadwich, 2012, p. 23).

1.2.2 Relationship to other marketing streams

The customer experience itself is not a new concept because it closely links to research streams within marketing. Among those are "customer satisfaction, service quality, relationship marketing, customer relationship management, customer centricity and customer engagement" (Lemon & Verhoef, 2016, p. 70). In their paper, Lemon and Verhoef (2016) identify seven developments which contributed to customer experience research in the past 60 years: first, customer buying behavior process models (1960s–1970s), second, customer satisfaction and loyalty (1970s), third, service quality research and customer journey mappings (1980s), fourth, relationship marketing in the 1990s, fifth, customer relationship management (CRM) and the elements of customer experience and their outcomes (2000s), sixth, the organizational challenges firms face with customer centricity when designing and managing the customer experiences (2000s–2010s), and lastly, customer engagement and the customer role in the experience (2010s) (Lemon & Verhoef, 2016, p. 71). Hwang and Seo (2016) also review the concept and delineate the term from concepts of customer satisfaction as a performance measure "as the outcome of the quality of the customer experience process" (p. 2220) and customer relationship management. They point out that a good customer experience leads to higher customer satisfaction. According to the researchers Hwang and SEO (2016), more empirical research on CEM should be carried out in the future.

It is beyond the scope of this chapter to explain all these concepts, since there is excellent literature available to review these particular marketing streams, for example, customer satisfaction, service quality, relationship management, customer centricity, and customer engagement as cited earlier.

1.3 Conceptualization of the customer experience

1.3.1 Conceptualization of the customer experience

There have been many attempts to conceptualize the customer experience or map the process of CEM. A selection of approaches is displayed in the following section.

Carbone and Haeckel (1994) propose four steps for designing an experience, for example, (1) acquisition of service experience design skills, (2) data collection and analysis, (3) service clue design, and (4) implementation. Service clue design is done with the service blueprinting method. Overall, experience design is a learning process for the firm with a typical learning curve.

In the early work of Schmitt (1999, p. 64), he proposes an experimental marketing framework with five experience modules: (1) sense, (2) feel, (3) think, (4) act, and (5) relate. The customer is supposed to have positive sensual experiences by using his or her five senses (sense), have affective experiences and experience moods and feelings (feel), cognitive experiences (think), physical experiences and changes in behavior (act), and lastly should identify with the product in cultural terms as part of a social group or consumer culture (relate). In 2003 he then proposed a five-step framework of how companies can create a customer experience: (1) analyzing the experiential world of the customer, (2) building the experiential platform, (3) designing the brand experience, (4) structuring the customer interface, and (5) engaging in continuous innovation (Schmitt, 2003, p. 25).

Berry and Carbone (2007, p. 31) proposed a five-step approach: (1) identify the emotions that evoke customer commitment, (2) establish an experience motif, (3) Inventory and evaluate experience clues, (4) determine the experience gap, and (5) close the experience gap and monitor execution.

A very comprehensive review shows that the conceptualization of the customer experience is multidimensional, involving several components such as cognition, emotion, behavior, senses, and social components (Bruhn & Hadwich, 2012; Lemon & Verhoef, 2016; Verhoef et al., 2009, pp. 5–36). Bruhn and Hadwich (2012, p. 13) identify six experience dimensions in their framework as displayed in the upper part of Fig. 1.1. The *sensory* dimension refers to the perception of environmental stimuli through the five senses. The *affective* dimension pertains to the emotional level of customers where moods or emotions are evoked as a reaction to the experience. The *cognitive* dimension refers to the consumer intellect. Experiences can initiate customer's intellectual stimulation and engagement in thought processes. The *behavioral* dimension relates to physical activities and interactions which are part of the experience. The *lifestyle* dimension delivers arguments for the confirmation of values and opinions of customers. The *social* is a relational dimension and relates to the interaction with others in a social context (Bruhn & Hadwich, 2012, p. 13).

All of these presented frameworks show that managing customer experience requires a critical and systematic situation analysis, an implementation plan, and a success evaluation.

1.3.2 Touchpoints on the customer journey

When a CEM strategy is chosen, firms need to understand the customer journey. Lemon and Verhoef (2016, p. 74) "conceptualize customer experience as a customer's 'journey' with a firm over time during the purchase cycle across multiple touch points." The authors call this a "dynamic process" (p. 74). This customer journey includes the customer's prepurchase or contact points with the firm before the product or service encounter like the assessment of expectations or past purchases as well as the evaluation of the experience itself at the time of purchase and after it has occurred (postpurchase) (Berry et al., 2002, p. 85; Lemon & Verhoef, 2016, pp. 74 and 76). Some of the external factors which occur during this customer journey are beyond the firm's control.

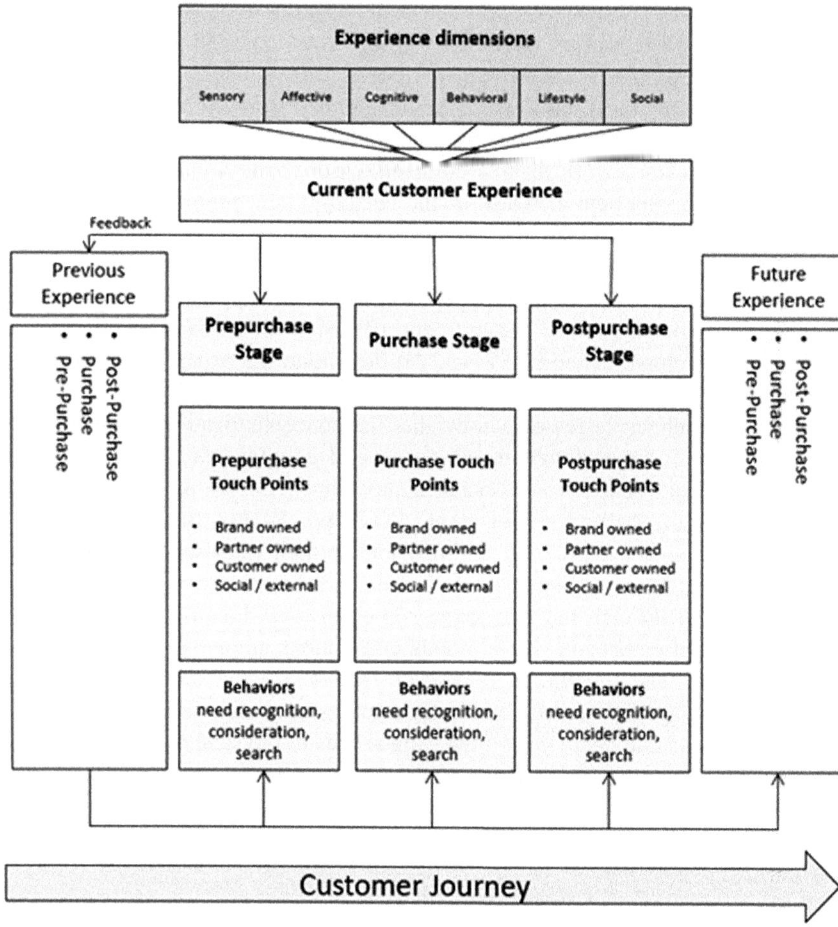

Figure 1.1 Process model for customer journey and experience and experience dimensions. *Source:* Own illustration adapted from Lemon, K. N., & Verhoef, P. C. (2016). Understanding customer experience throughout the customer journey. Journal of Marketing, *80*(November), 69–96. https://doi.org/10.1509/jm.15.0420 and Bruhn, M., & Hadwich, K. (2012). Customer experience – Eine Einführung in die theoretischen und praktischen Problemstellungen. In M. Bruhn, & K. Hadwich (Eds.), Customer experience - Forum Dienstleistungsmanagement (pp. 3–36). Springer Gabler. https://doi.org/10.1007/978-3-8349-4001-8_1.

Within the customer journey, Baxendale, Macdonald, and Wilson (2015) show that different customer touchpoints can be identified. These touchpoints should not be measured separately like in the past because the assessment of their relative impact is important for retailers to allocate their resources across touchpoints effectively. The importance of these touchpoints is individual and situation specific (Lemon & Verhoef, 2016, p. 76).

To capture the types of experience customers have on their customer journey in the context of shopping centers, Steffen (2012, p. 343) collected 443 customer experiences with the critical incident technique in and categorized these into 13 categories: assortment, atmosphere, other customers, price, store/center, feeling/well-being, weather/temperature, consumer specific, sales person, gift/samples, parking, numerous, and other. She developed a conceptual framework and measured emotional reactions and behavioral consequences of these customer experiences. A hierarchical structure model was built and tested which included 25 emotion descriptors and 8 behavioral outcomes (loyalty/disloyalty, short-term return/avoidance, affective reaction, complaining/word-of-mouth, do nothing, change in behavior/belief, more than one reaction, and other) (Steffen, 2012, pp. 394–398).

During the customer experience journey as displayed in the middle and lower parts of Fig. 1.1, the customer engages with different touchpoint categories (based on Lemon & Verhoef, 2016, p. 77). *Brand-owned* touchpoints are controlled by the firm, for example, through the marketing mix. *Partner-owned* touchpoints are controlled by the firm's partners such as marketing agency and loyalty program partners. *Customer-owned* touchpoints refer to experiences the customer is in control of, for example, product introductions or unboxing videos on Youtube. *Social/external* touchpoints point influence the customer, but these are mostly peer-related (e.g., touchpoints with the brand through other customers, bloggers, influencers in general, and reviews; Lemon & Verhoef, 2016, pp. 76–78).

The prepurchase, purchase and postpurchase stages from previous experiences are stored in the memory and the customer relates to them in the current experience situation. Furthermore how future experience is perceived depends on the experience a customer has made with the brand, product, or service partner in the past.

Firms should thus conduct a customer journey analysis to understand customer's options and choices and particularly how customers interact with the firm on all touchpoints in the three consumption phases. Service blueprinting and mobile research from a multichannel are proposed as ways to get insights on the customer journey (Lemon & Verhoef, 2016).

1.4 Firms challenges of creating experiences

The Marketing Science Institute sees a particular challenge in researching the customer experience in the upcoming years (Lemon & Verhoef, 2016). The future challenge for firms in creating a customer experience is that technological developments change the way firms offer experience. Virtual reality, mobile internet, and social media change the way people live and they are changing the way firms offer experiences. For example, rather than to visit museums or sights, customers can experience the exhibits online (Smit & Melissen, 2018, p. 2).

As noted earlier, many firms will have to change their core product and offer a total experience and redesign their business processes. The company Threadless, for example, started as a traditional t-shirt company in 2000 and is now selling shirts, accessories, home décor, apparel, and footwear canvases as a crowdsource business. The designs are submitted by artists around the world who compete in a voting system against each other to win a small price and the prestige to have their designs produced and offered online (Threadless, 2019).

Another example of a business which is redesigned for experience creation is Build-a-Bear workshop, a retail shop chain in which kids can create their own teddy bear in a series of six steps (e.g., choose a bear, listen to the bear, implant a heart, pick clothes, do a selfie, and get a certificate). The kids can individualize their bear, pick the fabric, and decide on how much stuffing they like (Bear, 2016). Smit and Melissen (2018, p. 3) point out that parents are prepared to spend more money on the bears because the process of customization offers an experience for their children.

A major challenge for firms will be to identify these process-changing experiences and design them to achieve a competitive advantage and to create a sustainable business model.

1.5 Outlook of the book

The book shows that consumers are becoming more critical in their food choices and the environmental impact of their food experiences. Although the food itself is not the only decision driver, the experience which accompanies the food is increasingly important to consumers.

The book is divided into four main parts. Part 1 includes Chapters 1–4, which review the literature on food choices and experiences and provide the theoretical background for the proceeding case-based chapters.

Chapter 1, Introduction to the Experience Economy, provided a brief introduction into the customer experience economy and briefly explained the customer experience concept and related theories.

In Chapter 2, Introduction to Consumer Food Choices, the consumer food choice process is explained by Pohlmann with reference to the Stimulus-Organism-Response model, Engel–Blackwell–Miniard model, and the Total Food Quality model. It furthermore proposes that consumer food choice criteria are changing because of new developments such as the clean label trend.

Chapter 3, The Role of Food in Creating Experiences, shows an overview about the role food plays in creating experience, for example, how food relates to memory and emotions. In addition, the editors Steffen and Doppler provide background information on food-related cultural and social phenomena. The chapter concludes with an implementation example of how marketers use food experiences for strategic destination branding and for event experiences.

By using the text-mining tool Leximancer, Capriello, and Hassan explore marketers unintended consequences of improving the consumer food experiences in Chapter 4, Marketers Unintended Consequences of Improving the Consumer Food Experience: Discovering the Key Drivers. The authors show that obesity, unhealthy eating, and nutrition are among the consequences in the Western world, especially for vulnerable marketing segments.

Part 2 consists of Chapters 5–7 and reviews food shopping experiences.

In Chapter 5, Crafting Food Shopping Experiences: The Case of Food Halls in Luxury Departments Stores, Eirini and Theodoridis demonstrate how food shopping experiences are crafted in luxury department stores around the world. They use the three dimensions (atmosphere, physical, and social) of retail experience to analyze food halls in Japan, France, United Kingdom, Germany, and Mexico.

Parves and his research team conduct a comparative cross-country study of organic food shopping between Australia and Bangladesh in Chapter 6, Organic Food Shopping in Australia and Bangladesh: The Role of Perceived Organic Food Value in the Attitude-Intention-Behavior Model. In particular this quantitative study shows that attitude and perceived organic food value are strong predictors of intentions and behavior. The chapter proposes that managers should use organic claims, attain trustworthy certifications, and educate stakeholders with a good communication mix.

Similarly, in Chapter 7, Promoting Organic Products in India: The Case of Two Brothers Organic Farms, Kumar, Singh, and Ambarkhane also investigate organic food marketing but in the Indian context. They identify the success factors of organic farming based on the case of Two Brothers Organic Farm. The results of the study show that the promotion of organic farming is especially successful when people are educated about organic farming. The two brothers are very enthusiastic and are prepared to take risks. They reach out to the wider community and engage in knowledge exchange, give lectures and workshops, encourage people to visit and stay on the farm, and build a very extensive distribution network.

Part 3 consists of Chapters 8–10 and illustrates cases on the food consumption experience in restaurants.

In Chapter 8, Engineering the Experience in Restaurants by Means of Process Design and Management, Hanson shows how gourmet restaurants in Brazil use business process management to engineer food experiences in restaurants. The case shows that many restaurants are still managed amateurish. They typically survive when professional business process management is applied and processes are clearly defined. Quality in food, service, and ambiance are usually responsible for long-term success.

Chapter 9, Ski Huts in Austria and Their Culinary Offer: A Question of Quality Versus Quantity?, provides an analysis of culinary food experience in ski huts in Austria. After conducting expert interviews with ski hut operators, Bauer-Krösbacher, Tischler, und Baumgartner show how important culinary experiences are for tourists. Although the logistics of providing a culinary food experience on a ski hut is challenging for the operators, the quality of the food, an innovative theme-based food presentation, and particularly the use of local products help to generate demand.

Bertella explores experience design strategies for vegetarian restaurants in Chapter 10, Food Experience Design for Plant-Based Restaurants. Similar to the findings in Chapter 8, Engineering the experience in restaurants by means of process design and management, and Chapter 9, Ski Huts in Austria and Their Culinary Offer: A Question of Quality Versus Quantity?, food quality and atmosphere are crucially important for providing customers a pleasant plant-based food experience. She however highlights the importance of competent staff, kitchen workers, and professional architects to provide pleasant experiences.

In Part 4 of the book, three cases on food experiences in the event and tourism context are illustrated in Chapters 11–14.

Doppler, Steffen, and Wurzer investigate how memorable holistic food experiences can be created by using the example Feinkost Käfer Event Catering in Germany. The focus of Chapter 11, Event Catering: Enhancing Customer Satisfaction by Creating Memorable Holistic Food Experiences, lies on identifying the factors that influence food experiences to enhance satisfaction. The findings also indicate that the quality of the food and ingredients, the quality of service in the preevent and live event phase, and the presentation of food and the fit between event design, atmosphere, and food are crucial factors for creating a memorable holistic food experience for customers and that especially unexpected elements in the experience create memorable experiences and therefore improve customer satisfaction.

Velasco Vizcaino, Pohlmann, and Cepeda investigate food experiences in personalized catering boutiques in Quito, Equador in Chapter 12, "We Have Our Style!" Optimizing Food Experiences in Personalized Catering Boutique Concepts. The results suggest that caterers should be open to customers' unusual ideas, cocreate value through customer participation, and tailor their offers to customers rather than offering recurring menus or dishes. A competitive advantage can be achieved by distinguishing between service and servility, by integrating and contextualizing new information, by facilitating social interactions, and by implementing entrepreneurial marketing to foster innovative capacities.

The contribution of Stovall, Hughes, and Ventzislavov in Chapter 13, Curated Nostalgia—Constructing and Consuming the City Through a Culinary Adventure, examines food as a city-defining tool in the scripted dinner series *Los Angeles Eats Itself.* Los Angeles Eats Itself is a curated, scripted event which not only refers to the dark episodes but also reexamines people's relationship with the city's identity.

In Chapter 14, Experiences Gained Through Absorption of and Immersion in Destinations' Food Events—Case Evidence From Finnish Travelers, Björk and Kauppinen-Räisanen explore the roles Finish travelers have at destination's food events. The authors investigate how important the basic human senses are for food experiences. The findings of the quantitative study show that experiences are gained through the absorption of and immersion in destinations' food events and that both provide experience value to travelers.

Chapter 15, Creating Food Experiences: Summary and Implications for Practitioners, summarizes the findings from all chapters and discusses the lessons practitioners can learn. Recommendations are presented for the following four themes, which emerged from the book chapters: information needs, strategic planning, strategy implementation, and unintended consequences.

The book concludes with an outlook. In Chapter 16, The Future of Food Experiences, Doppler and Steffen explore future developments in the context of food shopping experiences. Megatrends are here seen as major drivers of future food experiences.

References

Bagdare, S., & Jain, R. (2013). Measuring retail customer experience. *International Journal of Retail and Distribution Management*, *41*(10), 790–804. Available from https://doi.org/10.1108/IJRDM-08-2012-0084.

Batat, W. (2019). *Experiential marketing – Consumer behaviour, customer experience and the 7Es*. Milton Park: Routledge.

Baxendale, S., Macdonald, E. K., & Wilson, H. N. (2015). The impact of different touchpoints on brand consideration. *Journal of Retailing*, *91*(2), 235–253. Available from https://doi.org/10.1016/j.jretai.2014.12.008.

Bear, B. A. (2016). *Build a bear workshop experience*. Available at: <https://www.buildabear.com/brand-about-story-experience.html> Accessed 09.07.19.

Berry, L. L., & Carbone, L. P. (2007). Build loyalty through experience management. *Quality Progress*, *40*(9), 26–32.

Berry, L. L., Carbone, L. P., & Haeckel, S. H. (2002). Managing the total customer experience. *MIT Sloan Management Review*, *43*(3), 84–89.

Bruhn, M., & Hadwich, K. (2012). Customer experience – Eine Einführung in die theoretischen und praktischen Problemstellungen. In M. Bruhn, & K. Hadwich (Eds.), *Customer experience - Forum Dienstleistungsmanagement* (pp. 3–36). Gabler: Springer. Available from https://doi.org/10.1007/978-3-8349-4001-8_1.

Carbone, L. P., & Haeckel, S. H. (1994). Engineering customer experiences. *Journal of Marketing Management*, *3*(3), 8–19.

Doppler, S., & Steffen, A. (2018). Die Rolle des involvements für die wahrgenommene Erlebnisqualität von B2B Veranstaltungsbesuchen - Erkenntnisse für die markeninszenierende Livekommunikation. In C. Zanger (Ed.), *Events und Marke - Stand und Perspektiven der Eventforschung* (pp. 205–226). Wiesbaden: SpringerGabler.

Grönroos, C. (2006). Adopting a service logic for marketing. *Marketing Theory*, *6*(3), 317–333. Available from https://doi.org/10.1177/1470593106066794.

Hill, R. P., & Gardner, M. P. (1986). The buying process: Effects of and on consumer mood states. *Advances in Consumer Research*, *14*, 408–410.

Hirschman, E. C., & Holbrook, M. B. (1982). Hedonic consumption: Emerging concepts, methods and propositions. *Journal of Marketing*, *46*(Summer), 92–101. Available from https://doi.org/10.2307/1251707.

Holbrook, M. B. (2000). The millennial consumer in the texts of our times: Experience and entertainment. *Journal of Macromarketing*, *20*(2), 178–192.

Holbrook, M. B. (2006). Consumption experience, customer value, and subjective personal introspection: An illustrative photographic essay. *Journal of Business Research*, *59*(6), 714–725. Available from https://doi.org/10.1016/j.jbusres.2006.01.008.

Hwang, J., & Seo, S. (2016). A critical review of research on customer experience management. *International Journal of Contemporary Hospitality Management*, *28*(10), 2218–2246. Available from https://doi.org/10.1108/IJCHM-04-2015-0192.

Jain, R., Aagja, J., & Bagdare, S. (2017). Customer experience – A review and research agenda. *Journal of Service Theory and Practice*, *27*(3), 642–662. Available from https://doi.org/10.1108/JSTP-03-2015-0064.

Jüttner, U., Schaffner, D., Windler, K., & Maklan, S. (2013). Customer service experiences: Developing and applying a sequential incident laddering technique. *European Journal of Marketing*, *47*(5), 738–768. Available from https://doi.org/10.1108/03090561311306769.

Kirchgeorg, M., & Ermer, B. (2012). Wahrnehmungswirkung von Messeständen als temporären Markenerlebniswelten. In M. Bruhn, & K. Hadwich (Eds.), *Customer experience - Forum Dienstleistungsmanagement* (pp. 251–272). Wiesbaden: Springer Gabler. Available from https://doi.org/10.1007/978-3-8349-4001-8_11.

Lemon, K. N., & Verhoef, P. C. (2016). Understanding customer experience throughout the customer journey. *Journal of Marketing*, *80*(November), 69–96. Available from https://doi.org/10.1509/jm.15.0420.

Mymms. (2019). *Mymms UK homepage*. Available at: <https://www.mymms.co.uk/> Accessed 09.07.19.

Nivea. (2018). *Nivea Creme fotodose 75 ml*. Available at: <https://www.nivea.de/shop/creme-40059002647870001.html> Accessed 09.07.19.

Pine, B. J., & Gilmore, J. H. (1998). Welcome to the experience economy. *Harvard Business Review*, *July–Aug*, 97–105.

Prahalad, C. K., & Ramaswamy, V. (2004). Co-creation experiences – The next practice in value creation. *Journal of Interactive Marketing*, *18*(3), 5–18.

Schmitt, B. (1999). *Experiential marketing: How to get customers to sense, feel, think, act, and relate to your company and brands*. New York: Free Press.

Schmitt, B. H. (2003). *Customer experience management*. Hoboken, NJ: John Wiley & Son, Inc.

Shaw, C., & Ivens, J. (2002). *Building great customer experiences*. London: Palgrave Macmillan UK. Available from https://doi.org/10.1057/9780230554719.

Smit, B., & Melissen, F. (2018). *Sustainable customer experience design – Co-creating experiences in events, tourism and hospitality*. Abingdon, Oxon: Routledge.

Steffen, A. (2012). *Critical shopping experience – Affective reactions and behavioural consequences*. Saarbrücken, Germany: Südwestdeutscher Verlag für Hochschulschriften.

Threadless. (2019). *Threatless about*. Available at: <https://www.threadless.com/info/about/ > Accessed 09.07.19.

Verhoef, P. C., Lemon, K. N., Parasuraman, A., Roggeveen, A., Tsiros, M., & Schlesinger, L. A. (2009). Customer experience creation: Determinants, dynamics and management strategies. *Journal of Retailing*, *85*(1), 31–41. Available from https://doi.org/10.1016/j.jretai.2008.11.001.

Introduction to Consumer Food Choices

2

Attila Pohlmann
Department of Marketing, Business School, Universidad San Francisco de Quito, Ecuador

This introductory chapter provides an overview of prevalent consumer food choice models that are used to describe and predict consumer demand. Current theoretical challenges are discussed and contemporary developments regarding the *clean label trend* are outlined. Labels such as "healthy," "free from artificial additives," and "organic agriculture" have gained increasing importance for consumers, strongly influence their buying decisions, and are consequently of importance to marketing managers (Asioli et al., 2017).

2.1 Consumer Food Choice Models

Considered superficially, food choice appears to be a relatively simple matter driven primarily by biology: we get hungry, we eat. Yet, the food choices that consumers make vary greatly in their complexity, carry a variety of meanings, and are influenced by a multitude of cultural and social factors (e.g., Beardsworth & Keil, 2002; Cronin, McCarthy, & Collins, 2013; Pendergrast, 2016; Ulusoy, 2015). On top of that, the interactions of psychological, personal, situational, and environmental factors that affect consumers' food choices are complex. To describe and predict consumer behavior, researchers have devised a number of stimulus-organism-response models.

Early choice models relied on theories of reasoned action and planned behavior, suggesting that choice behaviors were rational and resulted from conscious processes, largely guided by economic considerations. Meanwhile, these models somewhat neglected psychological and environmental factors and many researchers suggest that most food decision-making occurs on a nonconscious, virtually automatic level (e.g. Aarts & Dijksterhuis, 2000; Adaval & Monroe, 2002; Ajzen & Fishbein, 2000; Grunert, 1996; Ji & Wood, 2007).

Although not specific to food choices, the Engel-Blackwell-Miniard (EBM) model (Engel, Blackwell, & Miniard, 1995) is a generalizable, multivariable consumer decision-making model that takes into account the effects of nonrational factors, such as environmental influences and personal differences on buying decisions. As depicted in Fig. 2.1, the core of the model is a cognitive problem-solving process of need satisfaction, yet the model accounts for influences of nonrational factors during many stages of the decision process. The EBM model indicates

Case Studies on Food Experiences in Marketing, Retail, and Events. DOI: https://doi.org/10.1016/B978-0-12-817792-1.00002-2

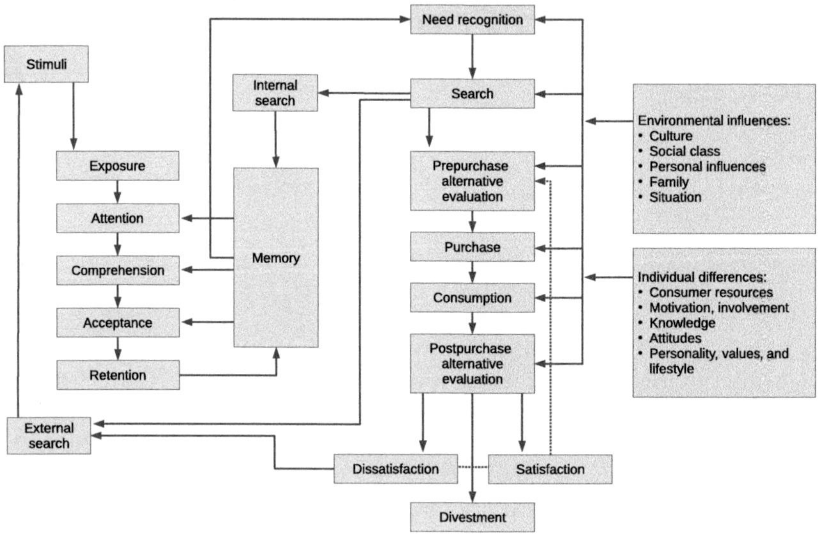

Figure 2.1 The Engel-Blackwell-Miniard model (Engel et al., 1995).

that nonrational factors can exert their influences throughout all stages of the decision-making process, starting with need recognition, all the way through product search, evaluation of prepurchase alternatives, the purchase itself, and finally consumption. Steenkamp (1997) adapts the EBM model specifically to food consumption by appending food properties, such as sensory perception and physiological effects as factors that influence the decision-making process.

Sensory properties of food products play a major role in consumer evaluation and preference formation. Therefore, food scientists scrutinize the chemical, physical, and sensory properties of food products, but their findings are—at least when viewed in isolation—rarely predictive of consumer choice and need to be situated in the entirety of the product's marketing mix (Garber, Hyatt, & Starr, 2003). One of the shortcomings of the aforementioned EBM model is that it does not take consumer perception into account. As an example, engineers and food technologists may rate a piece of beef as being of prime quality using objective criteria. However, some consumers may not have the ability, knowledge, or motivation to evaluate the beef's quality according to the same criteria and consequently perceive its quality in a different way. Even worse, they could end up ruining the beef in a failed home-cooking attempt, which will likely spell the end for the possibility of a positive evaluation of its quality.

Integrating these considerations, the widely cited Total Food Quality Model (TFQM, Grunert, 2005) describes how consumers form evaluations of food quality and safety. It is based on the means-end approach to consumer behavior (Reynolds & Olson, 2001), which suggests that consumers purchase food products to attain self-relevant life goals. Naturally, these goals can vary widely across segments and for individuals. According to the TFQM shown in Fig. 2.2, consumers assess food quality at two different points in time: (1) before the purchase when they form quality expectations relying on the quality cues that are communicated by the product's appearance and/or packaging and (2) after the purchase, basing their assessment on the experience with the product.

Before the purchase, to ensure that the product will meet their quality expectations, consumers decode and interpret the intrinsic and extrinsic quality cues that the product communicates. For instance, observable product attributes such as the color and fat content of a piece of meat are often used to infer taste and tenderness before the purchase. Consumer perceptions of quality tend to revolve around taste (and other sensory characteristics, such as smell), healthiness of the product, convenience, and increasingly process characteristics, such as organic certification, animal welfare considerations, and genetic modification (Brunsø, Fjord, & Grunert, 2002). Such credence attributes are communicated via packaging, certificates, labels, as well as by brand reputation and considerably contribute to the formation of consumers' quality expectations.

After the purchase, quality expectations are compared with experienced quality and result in a judgment on whether the actual experience lived up to expectations and whether it fulfilled the purchase motive. If so, a positive comparison between expectation and experience is likely to result in repeat purchases. Intervening factors that are largely out of the producer's control are the site of household production and consumption situation: an inspired hobby chef can spice up boring ingredients and combine them into a memorable, five-star, home-cooked meal; the efforts are likely to result in a positive evaluation. But even the finest of meals is difficult to appreciate when devoured in a rush to make it to the next appointment on time or when consumed in the company of cheerless guests, ultimately bearing down on purchase motive fulfillment evaluations.

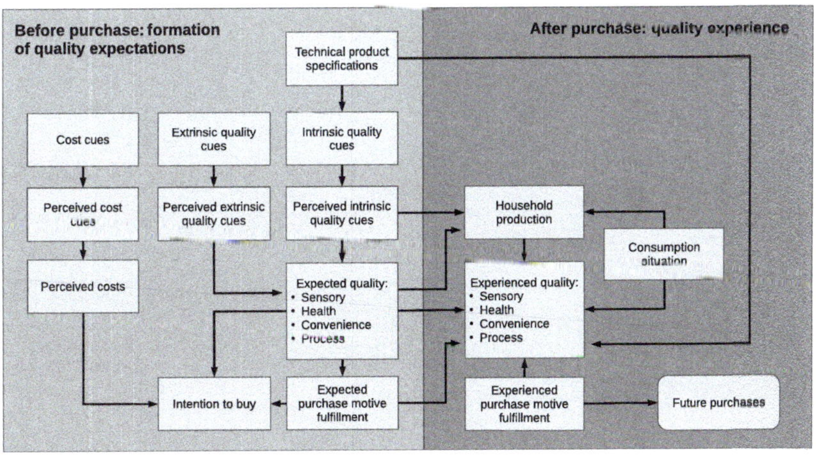

Figure 2.2 The Total Food Quality Model (Grunert, 2005).

Related to food quality, food safety is of major importance for public policy, but —due to its self-evidence—generally not a primary decision criterion for consumers in industrialized nations. However, food safety can become the overriding choice factor during times of crises (Grunert, 2005). Disease outbreaks, such as several cases of mad cow disease (CNN Library, 2019), listeria (The Guardian, 2018), as well as food contamination scandals, such as mislabeled "100% beef" horsemeat lasagna (BBC News, 2013) or high dioxin levels in animal feed (BBC News, 2011) render consumers' quality considerations secondary, often leading them to abstain from certain foods and entire categories altogether—at least for a while.

2.2 Current Challenges

After decades of research into the various factors that influence consumer food choices there is still no single, broadly accepted model that comprehensively explains consumer food choices (Marreiros & Ness, 2009). In view of the complexity of the aforementioned food choice models—and various other stimulus-organism-response models—Jacoby (2002, p. 52) remarks that they can take on the "intimidating appearance of a pipefitter's nightmare," at the same time questioning the usefulness of linear depictions of what rather appear to be fluid, psychological processes with various feedback loops among various systems.

Addressing these challenges, Köster (2009) calls for networked and interdisciplinary research approaches and proposes a framework with the aim to integrate efforts in psychology, biology, sociology, economics, marketing, and consumer research. Fig. 2.3 shows this model's six essential factors pertaining to food choice:

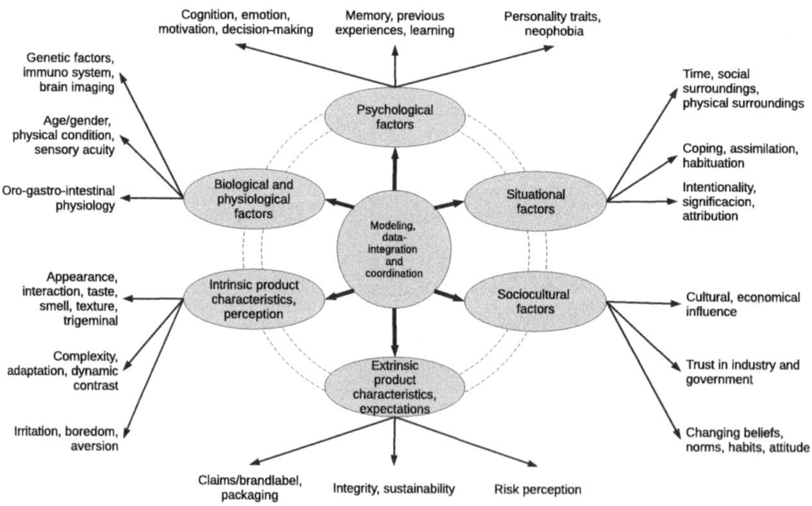

Figure 2.3 Interdisciplinary research framework with essential factors that influence food choice (Köster, 2009).

(1) perception of intrinsic product characteristics, (2) biological and physiological factors, (3) psychological factors, (4) situational factors, (5) sociocultural factors, and (6) expectations about extrinsic product characteristics.

Calling on psychological theories of learning, Köster (2009) suggests that many human behaviors are not consciously learned by solving repeat choice optimization problems, but are instead acquired through lifelong learning by way of imitation and social or environmental feedback, thus resulting in unconscious habit formation (Ji & Wood, 2007). Expanding on Grunert's (2005) notion of individual differences in perception and their influence on food choice, Köster (2009) raises the issue that objectively identical sensory stimuli can have vastly different hedonic valences depending on cultural context [e.g., acquired tastes such as Durian (large, oval, tropical fruit, fleshy on the inside with sharp points on the outside; common in South East Asia; produces a powerful smell, often likened to a mix of sulfur, onions, garlic, and cheese), Nattō (fermented and steamed soybeans with a stringy, glutinous texture, common in Japan and has a strong pungent smell), or Garnatálg (meat dish made with offal, covered in a layer of fat and traditional dish specific to the Faroe Islands)]. These observations raise questions about the prevalent assumptions regarding the cause—effect relationship between sensory stimulus and the learned post-ingestion satisfaction derived from it.

2.3 The Influence of Nonrational Factors

Many of the aforementioned nonrational factors may appear peripheral to the core of the decision-making system but can in fact have disproportionate effects on food choice and consumption behavior. In this context, it is important to distinguish between the different orientations of consumers, food producers, and marketers. For instance, the scope of the EBM model does not extend far beyond the purchase event and into the home of the consumer. This is most likely due to the fact that producers and marketers traditionally have little interest in what happens to their food products after the purchase. The TFQM goes further by encompassing the incalculable factors of household production and varying consumption situations, which are largely outside the sphere of influence of the producer anyhow. From a production and marketing perspective, models that predict purchase intention are certainly the most useful, especially for restaurants where consumption is relatively immediate and where the food preparation and consumption situation are under control. Meanwhile, not every item that is purchased in a store or ordered in a restaurant is actually consumed—potentially contributing to the problem of food waste (e.g., Aschemann-Witzel, de Hooge, & Normann, 2016; Block et al., 2016; Gruber, Holweg, & Teller, 2016; Porpino, Wansink, & Parente, 2016). On the flip side, for a variety of reasons, consumers often end up purchasing and consuming more than they initially intended, which can have implications for individual financial and physical health (Allan, Johnston, & Campbell, 2010; Ji & Wood, 2007).

Consumer behavior research articulates the sway that nonrational factors hold over food choices: negative affective states can increase overall food intake and preference for unhealthy foods (Garg & Lerner, 2013; Garg, Wansink, & Inman, 2007), whereas positive moods can promote healthier choices and reduce preference for indulgent foods (Gardner, Wansink, Kim, & Park, 2014). Additionally, the mere presence and physical appearance of others can profoundly influence consumption. McFerran, Dahl, Fitzsimons, and Morales (2010a) demonstrate that consumers anchor on the food quantities of the people around them and adjust these quantities according to the body size and shape of the other person. Furthermore, the goals of dieters and restrained eaters can be frustrated by the heavyweight appearance of their server (McFerran, Dahl, Fitzsimons, & Morales, 2010b), as well as by the designedly reduced packaging sizes of snacks, which are in fact intended to facilitate caloric restriction (Scott, Nowlis, Mandel, & Morales, 2008).

Furthermore, food choices often occur in an environment of hyperchoice, which is characterized by information overload (Mick, Broniarczyk, & Haidt, 2004). As a consequence—aside from relying on previous choices or habit—consumers often use heuristics in their decision-making processes since parsing through every bit of available information, product attribute, and product claim to arrive at an optimal choice is daunting—if not impossible. Depending on their individual food consumption goals (optimization, regulation, gratification, and necessity), consumers use vastly different heuristic strategies to gather goal-relevant information from nutritional labels (Chalamon & Nabec, 2016). With the intent to facilitate consumer informedness, a number of countries have introduced simplified *traffic light* nutrition labels. Suggesting that the brand itself can become a heuristic, recent research demonstrates that nutritional warning labels easily slip past consumers' attention nonetheless, especially when they are purchasing familiar brands (Velasco Vizcaíno & Velasco, 2019).

2.4 The Clean Label Trend

Over the course of the past decades, consumers have become increasingly interested in the production methods and the ingredients of the foods that they purchase (Asioli et al., 2017). This demand has led producers to communicate relevant information to consumers who perceive some production methods as more healthy and familiar than others, leading them to increasingly base their food-related decisions on such considerations. This so-called *clean label trend* can be attributed to rising consumer insecurity, which results from the increasing geographical and psychological distance between the sites of food production and its consumption (Princen, 1997; Weis & Weis, 2007), as well as from concerns about individual health and global sustainability (Grunert, 2013). Thus, with the widespread availability, relatively homogeneous quality, and affordability of most foods in industrialized nations, economic and sensory factors recede into the background when it comes to food choices. Instead, consumers attribute considerable importance to the information that is contained in clean labels. Surveys show that consumers try to avoid food products with unfamiliar ingredients, artificial additives, or unnatural production methods (Moskowitz, Beckley, & Resurreccion, 2012) and that they prefer natural and organic labeled foods instead (Euromonitor International, 2016; Katz & Williams, 2011).

These developments bring new challenges to research efforts to predict consumer behavior. Consumers' motivations as to whether and how to cognitively process clean label information can vary considerably among individuals, just as well as the decisions that are based on such information. Despite decades of research, to date, a commonly agreed upon model to describe and predict consumers' food choices still eludes researchers. The models that are introduced in this chapter are useful to conceptualize food choices and illustrate their complexity. To tackle their inherent complexity, more interdisciplinary research efforts are required.

2.5 Managerial Implications

Aside from economic and rational factors, nonrational factors greatly influence food choices, in most instances on an unconscious level. Due to the repetitive nature of food purchase and consumption behaviors, habits are formed and heuristics are used, which often take the place of more effortful, conscious decision-making. Once habits have formed, they are difficult to break, whether it is the individual who wishes to make dietary adjustments or the (social) marketer who wishes to promote specific behaviors. The goals of marketers and consumers can sometimes be at odds, but as consumers become more demanding, discerning, and informed about their food's origin and its production methods, the reconciliation of these conflicts presents great opportunities for both consumers and marketers.

References

Aarts, H., & Dijksterhuis, A. (2000). Habits as knowledge structures: Automaticity in goal-directed behavior. *Journal of Personality and Social Psychology*, *78*(1), 53–63.

Adaval, R., & Monroe, K. B. (2002). Automatic construction and use of contextual information for product and price evaluations. *Journal of Consumer Research*, *28*(4), 572–588.

Ajzen, I., & Fishbein, M. (2000). Attitudes and the attitude-behavior relation: Reasoned and automatic processes. *European Review of Social Psychology*, *11*(1), 1–33.

Allan, J. L., Johnston, M., & Campbell, N. (2010). Unintentional eating. What determines goal-incongruent chocolate consumption? *Appetite*, *54*(2), 422–425.

Aschemann-Witzel, J., de Hooge, I., & Normann, A. (2016). Consumer-related food waste: Role of food marketing and retailers and potential for action. *Journal of International Food & Agribusiness Marketing*, *28*(3), 271–285.

Asioli, D., Aschemann-Witzel, J., Caputo, V., Vecchio, R., Annunziata, A., Næs, T., & Varela, P. (2017). Making sense of the "clean label" trends: A review of consumer food choice behavior and discussion of industry implications. *Food Research International*, *99*, 58–71.

BBC News. (2011). Germany dioxins: Officials probe "illegal activity". *BBC News*. Retrieved April 2, 2019, from <https://www.bbc.com/news/world-europe-12134225>.

BBC News. (2013). Findus beef lasagne contained up to 100% horsemeat, FSA says. *BBC News*. Retrieved April 2, 2019, from <https://www.bbc.com/news/uk-21375594>.

Beardsworth, A., & Keil, T. (2002). *Sociology on the menu: An invitation to the study of food and society*. Routledge.

Block, L. G., Keller, P. A., Vallen, B., Williamson, S., Birau, M. M., Grinstein, A., ... Moore, E. S. (2016). The squander sequence: Understanding food waste at each stage of the consumer decision-making process. *Journal of Public Policy & Marketing*, *35*(2), 292–304.

Brunsø, K., Fjord, T. A., & Grunert, K. G. (2002). *Consumers' food choice and quality perception*. Aarhus, Denmark: The Aarhus School of Business Publ.

Chalamon, I., & Nabec, L. (2016). Why do we read on-pack nutrition information so differently? A typology of reading heuristics based on food consumption goals. *Journal of Consumer Affairs*, *50*(2), 403–429.

CNN Library. (2019). Mad cow disease fast facts. *CNN*. Retrieved April 2, 2019, from <https://edition.cnn.com/2013/07/02/health/mad-cow-disease-fast-facts/index.html>.

Cronin, J. M., McCarthy, M. B., & Collins, A. M. (2013). Covert distinction: How hipsters practice food-based resistance strategies in the production of identity. *Consumption Markets & Culture*, *17*(1), 2–28. <https://doi.Org/10.1080/10253866.2012.678785>.

Engel, J., Blackwell, R., & Miniard, P. (1995). *Consumer behavior*. Fort Worth, TX: Dryden Press.

Euromonitor International. (2016). *Report extract: Lifestyle 2016: New survey insights and system refresher*. London: Euromonitor International.

Garber, L. L., Hyatt, E. M., & Starr, R. G. (2003). Measuring consumer response to food products. *Food Quality and Preference*, *14*(1), 3–15.

Gardner, M. P., Wansink, B., Kim, J., & Park, S.-B. (2014). Better moods for better eating? How mood influences food choice. *Journal of Consumer Psychology*, *24*(3), 320–335.

Garg, N., & Lerner, J. S. (2013). Sadness and consumption. *Journal of Consumer Psychology*, *23*(1), 106–113.

Garg, N., Wansink, B., & Inman, J. J. (2007). The influence of incidental affect on consumers' food intake. *Journal of Marketing*, *71*(1), 194–206.

Gruber, V., Holweg, C., & Teller, C. (2016). What a waste! Exploring the human reality of food waste from the store manager's perspective. *Journal of Public Policy & Marketing*, *35*(1), 3–25.

Grunert, K. G. (1996). Automatic and strategic processes in advertising effects. *Journal of Marketing*, *60*(4), 88–101.

Grunert, K. G. (2005). Food quality and safety: Consumer perception and demand. *European Review of Agricultural Economics*, *32*(3), 369–391.

Grunert, K. G. (2013). *Trends in food choice and nutrition. Consumer attitudes to food quality products* (pp. 23–30). Wageningen: Wageningen Academic Publishers.

Jacoby, J. (2002). Stimulus-organism-response reconsidered: An evolutionary step in modeling (consumer) behavior. *Journal of Consumer Psychology*, *12*(1), 51–57.

Ji, M. F., & Wood, W. (2007). Purchase and consumption habits: Not necessarily what you intend. *Journal of Consumer Psychology*, *17*(4), 261–276.

Katz, B., & Williams, L. A. (2011). Cleaning up processed foods. *Food Technology*, *65*(12).

Köster, E. P. (2009). Diversity in the determinants of food choice: A psychological perspective. *Food Quality and Preference*, *20*(2), 70–82.

Marreiros, C., & Ness, M. (2009). A conceptual framework of consumer food choice behaviour. *CEFAGE-UE working papers*.

McFerran, B., Dahl, D. W., Fitzsimons, G. J., & Morales, A. C. (2010a). I'll have what she's having: Effects of social influence and body type on the food choices of others. *Journal of Consumer Research*, *36*(6), 915–929.

McFerran, B., Dahl, D. W., Fitzsimons, G. J., & Morales, A. C. (2010b). Might an overweight waitress make you eat more? How the body type of others is sufficient to alter our food consumption. *Journal of Consumer Psychology*, *20*(2), 146–151.

Mick, D. G., Broniarczyk, S. M., & Haidt, J. (2004). Choose, choose, choose, choose, choose, choose, choose: Emerging and prospective research on the deleterious effects of living in consumer hyperchoice. *Journal of Business Ethics*, *52*(2), 207–211.

Moskowitz, H. R., Beckley, J. H., & Resurreccion, A. V. A. (2012). *Sensory and consumer research in food product design and development*. John Wiley & Sons.

Pendergrast, N. (2016). Environmental concerns and the mainstreaming of veganism. In T. Raphaely, & D. Marinova (Eds.), *Impact of meat consumption on health and environmental sustainability* (pp. 106–122). Hershey, PA: IGI Global.

Porpino, G., Wansink, B., & Parente, J. (2016). Wasted positive intentions: The role of affection and abundance on household food waste. *Journal of Food Products Marketing*, *22* (7), 733–751.

Princen, T. (1997). The shading and distancing of commerce: When internalization is not enough. *Ecological Economics*, *20*(3), 235–253.

Reynolds, T. J., & Olson, J. C. (Eds.), (2001). *Understanding consumer decision making: The means-end approach to marketing and advertising strategy*. Mahwah, NJ: Lawrence Erlbaum, Psychology Press.

Scott, M. L., Nowlis, S. M., Mandel, N., & Morales, A. C. (2008). The effects of reduced food size and package size on the consumption behavior of restrained and unrestrained eaters. *Journal of Consumer Research*, *35*(3), 391–405.

Steenkamp, J.-B. E. M. (1997). *Dynamics in consumer behavior with respect to agricultural and food products. Agricultural marketing and consumer behavior in a changing world* (pp. 143–188). Boston, MA: Springer US.

The Guardian. (2018). Australians told to throw out rockmelon after two deaths in listeria outbreak. *Australia News. The Guardian*. <https://www.theguardian.com/australia-news/2018/mar/01/australians-told-to-throw-out-rockmelon-after-two-deaths-in-listeria-outbreak>.

Ulusoy, E. (2015). I think, therefore I am vegan: Veganism, ethics, and social justice. In *Annual macromarketing conference* (Vol. 419).

Velasco Vizcaíno, F., & Velasco, A. (2019). The battle between brands and nutritional labels: How brand familiarity decreases consumers' alertness toward traffic light nutritional labels. *Journal of Business Research*, *101*, 637–650.

Weis, A. J., & Weis, T. (2007). *The global food economy: The battle for the future of farming*. Zed Books.

The role of food in creating experiences

3

Adrienne Steffen and Susanne Doppler
University of Applied Sciences Fresenius Heidelberg, Heidelberg, Germany

3.1 Introduction to food experiences

Based on Maslow's hierarchy of needs, food is a basic need to survive (Solomon, Bamossy, Askegaard, & Hogg, 2016, p. 270). Marshall (2005, p. 2) characterizes food as "mundane," "essential," and "fuel" to then point out that we "literally consume (eat) the product, it comes part of us, and what we don't use becomes excrement." Furthermore Marshall (2005) pinpoints that food is highly individualistic, linked to social position and consumed for what it means to people instead of for pure nutrition. Maslow's pyramid however also shows that people in western cultures with a good economic situation shift to higher level needs and can obtain "peak experiences" (Solomon et al., 2016, pp. 270–271).

People have different priorities and generally enjoy eating food. But experiencing a dinner in a fancy restaurant, sampling fresh cheese at a market in Gouda, getting groceries in Harrods in London or visiting the pizza festival in Naples can be unique experiences which are remembered a lifetime. Experiences of this kind clearly do not classify as basic needs but address higher level needs. According to Smit and Melissen (2018, p. 48) these higher level needs could be esteem-needs and self actualization needs. Memorable experiences, which are admired by others, can fulfill social needs and extraordinary experiences which transform people can help with self-developing (Smit & Melissen, 2018, p. 48). Furthermore food is essential in social processes as a ritual but also as a gift exchange. In addition, food creates "continuity with the past" and builds memories (Sutton, 2008, p. 160).

In the retail shopping context, scholars differentiate between utilitarian (functional) and hedonic (pleasurable) shopping experiences (Solomon et al., 2016, p. 81). Utilitarian shopping takes place when consumers shop daily necessities. This type of shopping lacks excitement and is, for example, done with a shopping list. Hedonic shopping, however, is done for pleasure. It involves treasure hunting and finding luxury bargains or self-gifts (Hirst & Tresidder, 2016, p. 91). When firms want to offer a food experience, they can persuade consumers by triggering their imagination, emotions, or hedonism (Carvalho de Rezende & Silva, 2014, p. 92).

In the food consumption context, there are various options when eating outside the home. First, this can be commercial places specialized in food such as restaurants, fast-food restaurants, and snack bars. Second, it can be noncommercial places

Case Studies on Food Experiences in Marketing, Retail, and Events. DOI: https://doi.org/10.1016/B978-0-12-817792-1.00003-4

such as eating at family's and friend's houses. Third, service providers which offer food, such as hotels and restaurants (Carvalho de Rezende & Silva, 2014).

The concept of "experience" is used differently in the literature, but "they all relate to events, situations or encounters that demand attention and involvement of people experiencing them and they all lead to some form of memories or learning in combination with specific emotions" (Smit & Melissen, 2018, p. 14). Particularly food-related experiences occur every day as part of people's routine and daily activities (Kauppinen-Räisänenet al., 2013).

Experiences related to foods also have a broader, social role. Sutton (2008, p. 159) identifies food as "a key mediator of social relationships, a symbol of identity and a marker of difference, whether defined by gender, class, race, or ethnicity." Marketers use several techniques to create food experiences.

Experience marketing is closely linked to the social and cultural movements in society, especially rituals and trends. The experiences should differ from and be an escape from everyday life (Belk, Wallendorf, & Sherry, 1989; Hirst & Tresidder, 2016, p. 81) to get away from daily chores such as cooking, washing, and doing the dishes. Marketers draw customers into "dreamscapes" by evoking their emotions and addressing their needs (Hirst & Tresidder, 2016, p. 78). Experiences such as attending an event, going to the movies, a family holiday trip, or special meals all stay in the consumer's memory and help build their personal biography (Hirst & Tresidder, 2016, p. 84).

3.2 Learning from and remembering food experiences

Scholars from various disciplines have studied how food relates to memory historically, culturally, or personally. Food and memory can be linked in various ways, for example, food and identity, food and nostalgia, food and tradition, food and rituals, social dietary changes over time, and remembering and forgetting through food (Holtzman, 2006, p. 365).

Many food experiences are stored in our memory. Kauppinen-Räisänen, Gummerus, and Lehtola (2013, p. 667) summarize the existing literature and conclude that people store meaningful food experiences for three reasons. First, because the memories affect attitudes. Second, the food experiences people remember to influence consumer behavior and are relieved when desired. Third, because these food experiences evoke nostalgia. For example, children often remember when grandma cooked their favorite dish or when their mother prepared a dish which they demanded. When transferring this idea to the purchase of goods or services, there are several occasions for which special goods become markers for our lives such as buying the first car or the first house (Hirst & Tresidder, 2016, p. 83). Even more trivial experiences such as eating junk-food can evoke pleasurable memories. The memories of the fast-food consumption experience and industrial chemicals can be linked to people's childhoods, where junk-food was consumed worry-free (Sutton, 2008).

In conclusion, food seems to be linked to memory compellingly, because eating involves a lot of sensuality particularly through our smell and taste senses and appeals to nostalgia (Holtzman, 2006, p. 373).

3.3 Food experience evoked by emotions

Experience marketing creates a sensual world which people consume intentionally. The sensational process begins with the firm's marketing and promotion activities and has its highest point at the stage of actual consumption. Even after the consumption, people will remember the experience, talk about it or reminisce. Some people will even reimagine the experience at a later stage in time, where they construct new details of the story or image what could have been (Herbst & Muisolik, 2016, pp. 13–14; Hirst & Tresidder, 2016, pp. 94–95).

A customer experience can create many positive and negative emotions (Batat, 2019, p. 25). Throughout the journey, consumers experience the following path for a restaurant experience (e.g., discovery, enthusiasm, excitement, doubt, anxiety, denial, disillusion, disappointment, discovery, and enthusiasm). In the end, these emotions strongly impact satisfaction. Firms make use of emotional marketing strategies like customer-oriented service and communicate with tools like storytelling to appeal to consumers' emotions (such as love, happiness, and admiration) before they make them engage in cognitive processes (Batat, 2019, p. 25).

Daydreaming and nostalgia can also help people to drift off into pleasurable thoughts. These can be very real, and products can be part of these daydreams. Consumers use their own history and even use nostalgia to "inform their consumer choices and practices" and reexperience their past (Hirst & Tresidder, 2016, p. 97). Nostalgic emotions help to escape from every life. "Nostalgia originates from the Greek words *nostos* (homecoming) and *algos* (ache); it first described the homesickness caused by geographical distance" (Vignolles & Pichon, 2014, p. 226). In the marketing literature, nostalgia is mainly linked to childhood memories or family ties (Holtzman, 2006). Vignolles and Pichon (2014) conducted an exploratory study to investigate nostalgia and food consumption. They found food nostalgia to be linked to six central themes (childhood, yearning, substitute, homesickness, special occasions, and rediscovery). Although nostalgia was seen as a negative emotion, the result of Vignolles and Pichon (2014) suggests that people perceive food nostalgia to be a positive emotion, despite the notion that it was initially thought to be negative.

Marketers design and stage the gastronomy experience to touch people's emotions and appeal to nostalgic moments. Restaurants offer dinners which bring back people into specific decades, for example, 1950s dinners. Here the whole experience, including the dish and service environment is designed to resemble the original setting (Hirst & Tresidder, 2016, p. 100).

Another type of experience gastronomy which takes place in Germany is knight feasts as a form of entertainment. Participants and waiters dress for the occasion, medieval music is played, and people eat without cutlery (Auerbach, no date). People will learn from these food experiences, and they are likely to remember these food experiences for a long time.

Nostalgia does not have to be only seen as an experience which is reexperienced emotionally. It can also be a yearning for learning about cultures or places which were never visited (Holtzman, 2006).

3.4 Food experiences and culture

3.4.1 Food as a symbol of culture and religion

Food is a visible cultural dimension. People attach symbols to food, for example, religious symbols like in the Christian faith, boiled, and colored eggs on Easter are a sign of resurrection. Researchers found that food itself as a symbol expresses consumers' values and self-image in the sense of a "self-concept-product-concept" (Akbari, Gholizadeh, & Zomorrodi, 2018, p. 119 referring to Belk, Bahn, & Mayer, 1982, p. 5).

Food is strongly embedded in our culture (Akbari et al., 2018; Berg & Sevón, 2014; Lucarelli & Olof Berg, 2011) and every culture has set ideas of how to eat "properly" (Marshall, 2005). Traditional foods that have been formed throughout time are transformed into a cultural heritage by being passed down from generation to generation (Almli, 2012, cited in Kocaman, 2018). They strengthen the "sense of belonging [of people in countries and local regions and have an important role to play] in ensuring cultural continuity" (Guerrero et al., 2009, cited in Kocaman, 2018, p. 10). Hirst and Tresidder (2016, p. 81) even argue that marketers can create food experience which help people find stability in a rootless society.

The cultural importance of food is a motivator for people to travel. For many people, food experiences are central on vacation. On a holiday trip, a tourist can encounter local culture and food experiences, which can be memorable experiences. Batat (2019, p. 147) proposes three stages which all connect to form the overall food experience: first, food experience in the planning process. This stage involves cognitive thinking and excitement when tourists learn about and anticipate local food culture and popular dishes (Batat, 2019, p. 148). Timeless images in brochures and on websites signpost the experience and travelers create an own understanding of the expected experience (Hirst & Tresidder, 2016, p. 80). In the second stage, a lot of emotions are triggered when the tourist experiences the food at the spot either in a restaurant or on a local market. Interaction with native food cultures can be physical, for example, through the plates, ingredients or animals offered but also through the symbolism of food, the social values which are attached to food, the artisanal production, etc. The tourist learns about table manners and food intake, the places of supply, the rules of hygiene, and distinguishes these from the boundaries of his own food culture. For the tourist "eating the food of the 'Other' [...] is an experiential learning process in itself" (Batat, 2019, p. 148). In many destinations, it is also possible to take cooking classes with locals, so they literally learn how to prepare and eat the dishes like locals do.

The last stage, the food memory stage, takes place after the experience happened. With their food memory, tourists stay connected to their experience. Often food products such as wines or spices are taken home and therefore strengthen and extend the trip. Many people also adopt new food practices and cooking or eating habits and share these with their friends (Batat, 2019, p. 148).

Food culture is thus a central element in our lives, which is learned through reading about other food cultures, through everyday experience or through cultural encounters when going on holidays. These food experiences are often linked to myth and rituals.

3.4.2 Food myth and food rituals

Rituals have symbolic action. In the context of cultural theories rituals are "the means by which the cultural system and the social system are able to interact and harmonize with each other" (Bell, 1997, p. 61). A more universal definition of rituals identifies a ritual as "how and why people produce ritualized actions; it certainly obscures one of the most decisive aspects of ritual as a strategic way of acting, the sheer degree of ritualization that is invoked" (Bell, 1997, p. 82). Thus rituals are complex sets of activities which construct meaning and values in specific ways (Bell, 1997, p. 82).

Sutton (2008) conducted ethnographic research on the Greek island of Kalymnos to study the preparation of Easter lambs, which takes place every year in self-build closed containers. The local people build the oven in preparation to end their fasting days and get lambs from family and friends or a local farmer because trust is essential. A lot of personal exchange and communication takes place during the preparation process. Kids play with the lambs before it is slaughtered, and while building the oven, the adults anticipate the good taste. These rituals are passed on from generation to generation (Sutton, 2008).

To demystify historical food rituals and satisfy their "nostalgic urge," tourists visit pilgrimage or heritage museums and taste heritage food (Hirst & Tresidder, 2016, p. 101). Eating this heritage food seems to be linked to our collective memory (Holtzman, 2006, p. 373). But on a daily basis, we take our food rituals, habits, and routines for granted (Chitakunye & Maclaran, 2008). Eating is typically a ritual involving the nuclear family. It is a time especially during holiday season when family and friends bond like on Easter or Christmas (Belk et al., 1989). Eating is a social activity where they meet and interact with each other. This counts for informal setting at home but also for eating out.

In private homes, the formal setting of the table, the dining order, and formalities done in each family create social order (Marshall, 2005). At home, people who work full-time jobs or families often struggle to find enough time to eat and juggle all chores. The "quality times" parents have with their children includes playing, vacation, and excursions and spending time at the dinner table (Chitakunye & Takhar, 2014, p. 1165). Although food appeared to be a social activity in the past, which was mainly done at home in the inner circle of the family, young consumers seem to adopt new eating practices. Young consumers adapt practices in informal settings such as eating in front of the television, eating at any time, and speed-eating (Chitakunye & Maclaran, 2008), mainly because this traditional dinner time does not take place because families cope with time pressure (Alm & Olsen, 2017).

Food rituals "act as scripts regulating the sequence and order of the dishes and provide guideline on appropriateness and comportment in certain situations" (Marshall, 2005, p. 4).

Scripts "constitute the knowledge of how to do things" (Shoemaker, 1996, p. 43), for example, following a recipe is an instrumental script or entering a fast-food restaurant is a situational script. All activities which involve a specific experience or event are part of the customer's script. These customer scripts are divided into subunits of activities or actions, for example, "read menu," "eat appetizer," and "hand server credit card" (Shoemaker, 1996, p. 44). There are various actors involved in each episode of the script and these scripts form customer's expectations of what will happen next in a service situation (Shoemaker, 1996, p. 49).

Food delivery apps help to extend the customer experience. They challenge traditional restaurants and delivery models. As a consequence of this technological disruption, restaurants try to offer a cozy atmosphere to get people to go to restaurants instead of getting the food delivered to their own home (Batat, 2019, p. 150).

3.4.3 Food experience and time

People increasingly feel time pressure and engage in time-saving consumption practices. Some meals are chosen spontaneously right at the retailer's shelves, others are carefully planned in advance. To save time, consumers do one-stop shopping and get the groceries for the week, whereas others either purchase convenience foods or rather eat out (Davies & Mandran, 1997). Alm and Olsen (2017, p. 105) found that Norwegian families cope with time pressure by "skipping dinner and eating snacks instead, consuming convenience food, avoiding preference conflicts, planning for healthy dinners, involving children and grandparents in food preparation, and practicing compensatory health beliefs and behaviors."

Despite all the convenience food which is offered in supermarkets, the research of Davies and Mandran (1997) found that in the United Kingdom, there is still a significant proportion of people who enjoy spending time in the kitchen to prepare food. They call this group people who are "willing to make time for food" and proposes that this rather traditional group of consumers prefers the experience of shopping in delicatessen and specialty stores.

Many consumers, however, perceive time pressure and turn to fast food or convenience food in supermarkets. The idea of fast food and supermarkets is, on the one hand, to increase consumer participation and on the other hand, to enhance efficiency and to reduce cost. This increase in participation should also improve the consumption experience and provide consumers with "satisfaction, values, and reward" (Batat, 2019, p. 132).

Even fast-food restaurants are now changing the way they utilize time and space. McDonalds has introduced an app with mobile order and pay and in-store order—terminals which allow customers to individualize their offers and make best use of their time (McDonalds, no date).

People find pleasure in different ways. Many spend a lot of time to go food shopping for special occasions, for example, to find unusual ingredients in different shops and on the local market and to cook a meal which requires a lot of time and leaves a mess to clean up in the kitchen. Alternatively they could go out to eat in a restaurant or they could order an individualized meal prepared by an expert to be delivered at home. According to Hirst and Tresidder (2016, p. 89), this experience of ordering, making changes, and eating is "elevated to a hedonistic aesthetic that underpins the concept of experience marketing."

The next section provides an implementation scenario of how food experiences are created in destination marketing and in the marketing of events.

3.5 Implementation scenarios: the role of food experience in destination marketing and marketing events

All over the world food is often used to brand cities and places (Berg & Sevón, 2014, Lucarelli & Olof Berg, 2011) and to develop destinations (Björk & Kauppinen-Räisänen, 2016; Raji, Karim, Arshad, & Ishak, 2018). Even in the "branding of national and ethnic identities [food plays an essential role] for example, in Italian, French or Thai Cuisines" (Berg & Sevón, 2014, p. 291). Consequently there is much literature about how food and gastronomy in various forms has been recognized as an essential element to make cities and places attractive destinations for visitors, the general public and businesses (e.g., Berg & Sevón, 2014; Lee, Packer, & Scott, 2015; Seo, Yun, & Kim, 2017; Zahari, Zain, & Hanafiah, 2018) analyzed and reviewed, for example, by the "Stockholm Program of Place Branding (STOPP)" (Stockholm Business School, no date). Lucarelli and Olof Berg (2011) and Berg and Sevón (2014) provide good insights into this field of research.

The approach of strategic destination marketing is "possibly the most conspicuous way in which food is used in city branding" (Berg & Sevón, 2014, p. 293), appearing as food or culinary tourism (Sohn & Yuan, 2013). Furthermore cities are branded by food festivals (Berg & Sevón, 2014), such as the Munich Oktoberfest, the Grüne Woche in Berlin, the Wurstmarkt in Bad-Dürkheim, on the whole German examples of iconic food festivals which have now become veritable hallmark events (Allen, O'Toole, Harris, & McDonnell, 2010, pp. 13–14). Berg and Sevón (2014, p. 294) summarize three categories of food-related city branding strategies: first, event-based strategies, for example, [food] festivals and exhibitions (e.g., López-Guzmán, Uribe Lotero, Pérez Gálvez, & Ríos Rivera, 2017), second, excellence and core competence strategies, for instance "City of Gastronomy and City of Origin" (e.g., UNESCO Cities of Gastronomy, Pearson & Pearson, 2016), third, various architectural and spatial strategies, for example, iconic food halls and waterfront restoration areas. Food-related strategic destination marketing seems to pursue the aims of "supporting food industry, protecting and amplifying identities of places and to change the place" (Berg & Sevón, 2014, p. 294).

Food experience plays an essential role in marketing communication, particularly in marketing events such as trade shows and other industry events as a specific tool in companies' marketing mix.

Food and thus catering offers at marketing events can be seen as an essential part of the staging of marketing events (Ali, Ferdinand, & Chidzey, 2017, pp. 75–76) Thus they fulfill their contribution to evoke event experiences, aiming to increase the positive effect on participants and other stakeholders (Orefice, 2018) related to the activation of sensory and emotional experiences (Ali et al., 2017, pp. 75–76). Consequently food offerings at events are one essential factor that "might impact directly on visitors' perceptions of overall service quality" (Gottlieb, Brown, & Drennan, 2011, pp. 1654) and are strongly related to experiences (Smit & Melissen, 2018, p. 24). As companies compete for the attention of their target groups, they strive to combine unique, memorable experiences with their products and services (Doppler, 2016, pp. 5–6), as experience substitutes quality as a competitive battlefield in marketing (Klaus & Maklan, 2013). Similarly to O'Sullivan and Spangler (1999) cited in Berridge (2007, p. 120), who explain main characteristics of experience as the emotional and cognitive undergo, which result in participation and involvement consequently. Involvement then is an important prerequisite for purchasing decisions (Kroeber-Riel & Gröppel-Klein, 2013, p. 463) and is therefore explicitly evoked by event managers at marketing events (Doppler, 2016, pp. 43–44).

Nevertheless, it would be too brief to understand the food experiences as a savior to trigger involvement since it is known, that the exchange and value cocreation among stakeholders over time is increasingly getting into marketers' focus (Brand & Rocchi, 2011), which also results in the way event professionals design and stage engaging experiences (Orefice, 2018). In this perspective, food experiences gain a new meaning at marketing events, in which it is less about the food consumption itself experienced, for example, in good taste and quality (Walker, 2005), but rather the social meaning of food experiences comes to the fore. Food offers a platform for social exchange and shared experience which is further discussed in Chapter 12, "We Have Our Style!" Optimizing Food Experiences in Personalized Catering Boutique Concepts. In addition, food is an expression of joint values, for example, the consumption of sustainably produced products and vegan lifestyle. More practical implications and the future of food experiences are discussed in more detail in Chapters 15 and 16.

3.6 Conclusions

Although food is a basic human need, most people see food shopping, food preparation, and food consumption as a way to satisfy higher level needs such as self-esteem or self-actualization needs. People have food-related experiences on a daily basis. Food habits and routines are embedded in our culture, and our society has norms of acceptable table manners. The way people use their time to eat, for example, whether dinner and food preparation is seen as quality time with the family or whether it is a stressful experience which is squeezed between two tasks by serving a ready-made meal is situation specific but also linked to people's values.

Marketers are designing and staging unique food experience to get people's attention. Companies compete for attention and strive to combine unique, memorable experiences with their products and services. Thus experience gastronomy and also event marketers aim to touch people emotionally through designing and staging a unique servicescape or by evoking memorable moments. Marketers, for example, use unique food cultures for destination branding. Many people travel, visit restaurants, or order exotic food to have unique food experiences.

References

Akbari, M., Gholizadeh, M. H., & Zomorrodi, M. (2018). Islamic symbols in food packaging and purchase intention of Muslim consumers. *Journal of Islamic Marketing*, *9*(1), 117–131.

Ali, N., Ferdinand, N., & Chidzey, M. (2017) Event design. In Ferdinand, N., Kitchin, & P.J. (Eds.), *Events management – An international approach* (pp. 67–93). London: Sage.

Allen, J., O'Toole, W., Harris, R., & McDonnell, I. (2010). *Festival and special events management* (5th ed.). Hoboken, NJ: Wiley & Sons.

Alm, S., & Olsen, S. O. (2017). Coping with time pressure and stress: Consequences for families food consumption. *Journal of Consumer Policy*, *40*(1), 105–123.

Auerbach, S. (no date). *Knights feast Castle Auerbach*. Available from <https://www.schloss-auerbach.de/> Accessed 25.03.19.

Batat, W. (2019). *Experiential marketing – Consumer behaviour, customer experience and the 7Es*. Milton Park: Routledge.

Belk, R. W., Bahn, K. D., & Mayer, R. N. (1982). Developmental recognition of consumption symbolism. *Journal of Consumer Research*, *9*(1), 4–17.

Belk, R. W., Wallendorf, M., & Sherry, J. F. J. (1989). The sacred and the profane in consumer behavior: Theodicy on the Odyssey. *Journal of Consumer Research*, *16*(1), 1–38.

Bell, C. (1997). *Rituals: Perspectives and dimensions*. New York: Oxford University Press.

Berg, P. O., & Sevón, G. (2014). Food-branding places – A sensory perspective. *Place Branding and Public Diplomacy*, *10*(4), 289–304.

Berridge, G., Events Design and Experience, 2007, Elsevier, Oxford.

Björk, P., & Kauppinen-Räisänen, H. (2016). Local food: A source for destination attraction. *International Journal of Contemporary Hospitality Management*, *28*(1), 177–194.

Brand, R., & Rocchi, S. (2011). Rethinking value in a changing landscape. A model for strategic reflection and business transformation. *A Philips design paper*. Available from <www.researchgate.net/publication/272941478_Rethinking_value_in_a_changing_landscape_A_model_for_strategic_reflection_and_business_transformation> Accessed 12.04.19.

Carvalho de Rezende, D., & Silva, M. A. R. (2014). Eating-out and experiential consumption: A typology of experience providers. *British Food Journal*, *116*(1), 91–103.

Chitakunye, D. P., & Maclaran, P. (2008). The everyday practices surrounding young peoples food consumption. *Young Consumers*, *9*(3), 215–227.

Chitakunye, P., & Takhar, A. (2014). Consuming family quality time: The role of technological devices at mealtimes. *British Food Journal*, *116*(7), 1162–1179.

Davies, G., & Mandran, C. (1997). Time, food shopping and food preparation: Some attitudinal linkages. *British Food Journal*, *99*(3), 80–88.

Doppler, S. (2016). *B2B eventmarketing* (1st ed.). München: UVK Verlagsgesellschaft.

Gottlieb, U. R., Brown, M. R., & Drennan, J. (2011). The influence of service quality and trade show effectiveness on post-show purchase intention. *European Journal of Marketing*, *45*(11/12), pp. 1642–1659.

Herbst, G., & Muisolik, T. H. (2016).). *Digital storytelling: Spannende geschichten für interne kommunikation, werbung und PR (PR praxis)*. Köln: Herbert von Halem Verlag.

Hirst, C., & Tresidder, R. (2016). *Marketing in food, hospitality, tourism and events – A critical approach* (2nd ed.). Woodeaton: Goodfellow Publisher.

Holtzman, J. D. (2006). Food and memory. *Annual Review of Anthropology*, *35*, 361–378.

Kauppinen-Räisänen, H., Gummerus, J., & Lehtola, K. (2013). Remembered eating experiences described by the self, place, food, context and time. *British Food Journal*, *115*(5), 666–685.

Klaus, P., & Maklan, S. (2013). Towards a better measure of customer experience. *International Journal of Market Research*, *55*(2), 227–246.

Kocaman, E. M. (2018). A cross-cultural comparison of the attitudes of employees towards the presence of traditional foods in business menus. *International Journal of Gastronomy and Food Science*, *13*.

Kroeber-Riel, W., & Gröppel-Klein, A. (2013). *Konsumentenverhalten*. München: Vahlen.

Lee, K. H., Packer, J., & Scott, N. (2015). Travel lifestyle preferences and destination activity choices of slow food members and non-members. *Tourism Management*, *46* (February), 1–10.

López-Guzmán, T., Uribe Lotero, C. P., Pérez Gálvez, J. C., & Ríos Rivera, I. (2017). Gastronomic festivals: Attitude, motivation and satisfaction of the tourist. *British Food Journal*, *119*(2), 267–283.

Lucarelli, A., & Olof Berg, P. (2011). City branding: A state-of-the-art review of the research domain. *Journal of Place Management and Development*, *4*(1), 9–27.

Marshall, D. (2005). Food as ritual, routine or convention. *Consumption Markets & Culture*, *8*(1), 65–85.

McDonalds. (no date). McDonalds mobile order and pay. Available from: <https://www.mcdonalds.com/us/en-us/mobile-order-and-pay.html> Accessed 18.03.19.

Orefice, C. (2018). Designing for events – A new perspective on event design. *International Journal of Event and Festival Management*, *9*(1), 20–33.

Pearson, D., & Pearson, T. (2016). Branding food culture: UNESCO creative cities of gastronomy. *Journal of International Food and Agribusiness Marketing*, 164–176.

Raji, M. N. A., Karim, S. A., Arshad, M. M., & Ishak, F. A. C. (2018). Community development through food tourism: Exploring the utilization of local food as community development at rural destination in Malaysia. *International Journal of Academic Research in Business and Social Sciences*, *8*(10), 937–951.

Seo, S., Yun, N., & Kim, O. Y. (2017). Destination food image and intention to eat destination foods: A view from Korea. *Current Issues in Tourism*, *20*(2), 135–156.

Shoemaker, S. (1996). Scripts: Precursor of consumer expectations. *Cornell Hotel and Restaurant Administration Quarterly*, *37*(1), 42–53.

Smit, B., & Melissen, F. (2018). *Sustainable customer experience design, sustainable customer experience design*. Oxon: Routledge.

Sohn, E., & Yuan, J. (2013). Who are the culinary tourists? An observation at a food and wine festival. *International Journal of Culture, Tourism, and Hospitality Research*, *7*(2), 118–131.

Solomon, M. R., Bamossy, G. J., Askegaard, S., & Hogg, M. K. (2016). *Consumer behaviour – A European perspective* (6th ed.). Harlow: Pearson.

Stockholm Business School . Stockholm Programme of Place Branding (STOPP). Available from: <https://www.sbs.su.se/forskning/marknadsf%C3%B6ring/forskningsprogram/2.44191/stockholm-programme-of-place-branding-stopp-1.274185> Accessed 30.06.20.

Sutton, D. (2008). A tale of Easter ovens: Food and collective memory. *Social Research*, *75* (1), 157–181.

Vignolles, A., & Pichon, P.-E. (2014). A taste of nostalgia. *Qualitative Market Research: An International Journal*, *17*(3), 225–238.

Walker, B. (January 2005). Foodservice – What the future holds. Caterer & Hotelkeeper (. 18).

Zahari, M., Zain, N., & Hanafiah, M. H. (2018). Food and tourism destination image: Moderating effect of local food consumption. *e-Review of Tourism Research*, *15*(1), 21–36.

4 Marketers' unintended consequences of improving the consumer food experience: discovering the key drivers

Antonella Capriello[1] *and Rohail Hassan*[2]

[1]Department of Economics and Business Studies, University of Piemonte Orientale (UPO), Novara, Italy, [2]Othman Yeop Abdullah Graduate School of Business (OYAGSB), Universiti Utara Malaysia (UUM), Sintok, Malaysia

4.1 Introduction

A noted attempt against fast food marketing strategies was the demonstration at a proposed McDonald's site near the Spanish Steps in Rome in 1986 (Suro, 1986). This event was fundamental in founding the Slow Food Movement whose philosophy is to prevent the disappearance of local food cultures and traditions and counteract the rise of fast life, proposing an alternative approach to food production and science based on respecting local foods (Slow Food, 2019).

Despite the slow movement, fast food restaurants hold a competitive position in the global marketplace, and the increasing number of such restaurants is a significant predictor of local obesity trends (Currie, Dellavigna, Moretti, & Pathania, 2010). Furthermore the diffusion of ultraprocessed foods is a key driver of the global food system associated with higher profits over fresh or minimally processed foods (Monteiro, Moubarac, Cannon, Ng, & Popkin, 2013).

As Fry and Polonsky (2004) underline, "while many firms engage in successful marketing activities with outcomes beneficial for both the firm and its stakeholders, a number of situations occur where these successful outcomes impact in an unanticipated negative fashion on consumers, society and other stakeholders," we relate this concept to the notion of the unintended consequences of improving the consumer food experience, negatively affecting consumer well-being in an unanticipated way. The prevailing pattern of food and beverage marketing aimed at children is a threat to the health of the next generation (Kraak, Gootman, & McGinnis, 2006). More specifically Buijzen and Valkenburg (2012) underscore that the effects of advertising on children may have unintended effects, such as generating a materialistic orientation, parent–child conflicts, and unhealthy eating habits.

Conversely creating an environment in which youngsters can grow up healthily should be a priority for nations. Hence the importance of analyzing and discussing the fundamental aims of food marketing approaches, particularly with regard to

Case Studies on Food Experiences in Marketing, Retail, and Events. DOI: https://doi.org/10.1016/B978-0-12-817792-1.00004-6

healthy aging policies in the Western world. Specifically the research focuses on the key drivers of the unintended consequences of improving the consumer food experience, reflecting on specific food categories and evolution of studies in the past two decades by adopting a content analysis using a text-mining tool for analyzing and visualizing the concepts and themes in the texts.

To build alternative future marketing approaches for a healthy ageing society, this study reflects on the need for essential and timely preventive nutrition strategies aimed at maintaining or improving the quality of life of the aging population (Kalache & Kickbusch, 1997).

This chapter is structured as follows: Section 4.2 describes the research design tracking the evolution of research related to 52 past studies. Section 4.3 presents the results of the specific food strategy traits, values, and orientations. Section 4.4 concludes by highlighting the importance of a new perspective to revitalize and advance the research field.

4.2 Research design

In this study, we conducted an automated content analysis and collected data from 52 articles published in journals including marketing and consumer research, the related social science disciplines, and nutrition research between the years 2000 and 2019. To identify the studies, a focus group with eight experts in food marketing was organized to determine the link between the customer food experience and the unintended consequences of the experience. The expert group identified the following keywords: *commercial food (or processed food)*, *junk food* (describing food that is high in calories from sugar or fat, with little dietary fiber, protein, vitamins, or minerals), *fast food* (food prepared and served quickly at fast food outlets), and *ultraprocessed food* (substances derived from food through baking, frying, extruding, molding, reshaping, hydrogenation, and hydrolysis, in accordance with Monteiro et al., 2013). These keywords informed the interrogation of websites and major indexing databases (i.e., *Web of Science* and *Scopus*). We selected articles based on the abstract in connection with consumer food experience and food marketing strategies. For an effective content analysis that shows how the concepts in consumer food experience and marketing strategies have developed from 2000 to 2019, we designed a subset of data in 5-year periods (2000–05, 2006–10, 2011–15, and 2016–19). Table 4.1 presents the distribution of selected research articles by years, clearly showing that the 2006–10 period was prominent in research in this field. The percentage of the literature refers to the number of articles considered in this study classified under the research topic in *Scopus* and *Web of Science* from 2000 to 2019.

Fig. 4.1 plots the total number of articles (empirical and other publications) published in the 5-year intervals. A cumulative total of 52 publications were identified for this study.

We used the Leximancer text-mining software to map the themes and concepts, using the visible concepts range to better understand the targeted data, similar to the study of Watson, Smith, and Watter (2005).

Table 4.1 Articles on the key drivers of marketers' unintended consequences of improving the consumer food experience.

Year	Number of articles	Percentage of the literature
2000–05	2	3.85
2006–10	19	36.54
2011–15	21	40.38
2016–19	10	19.23
Total	52	100

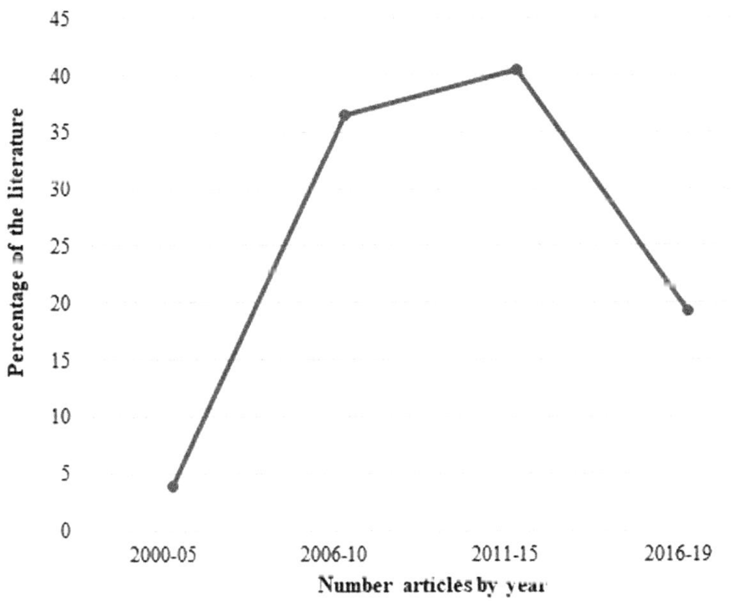

Figure 4.1 Articles about marketers' unintended consequences of improving the consumer food experience.

4.2.1 *Leximancer analysis*

We used the Leximancer (ver. 4.5) software for automated content analysis, which has a high level of proximity values for text mining and natural language processing to automatically classify concepts in textual data and map the themes (Smith & Humphreys, 2006). According to Bayesian theory, the texts can be quantified by identifying word frequency as concepts and their associations through an unsupervised process (Smith & Humphreys, 2006). According to Watson et al. (2005, p. 1233), "fragmented pieces of evidence" in a dataset "can be employed to predict what is happening in a system." Angus, Rintel, and Wiles (2013) explain that

Leximancer outputs in the form of visual demonstrations and themes can be color coded, with brightness showing the prominence of themes. The concepts have a strong semantic relationship if mapped closely to each other (Campbell, Pitt, Parent, & Berthon, 2011). Leximancer software has an interactive concept-mapping feature that helps researchers interpret the conceptual structure. Another essential point is that the theme is based on the most prominent concept in a specific group, even if the themes can be renamed where necessary (although in this study, we did not rename any). Researchers can use concept maps, text query options, and rank-ordered lists to understand specific concepts and their relationships. In brief, the visual map presents the strength of association among concepts and provides a conceptual overview of the data. We used Leximancer to analyze the entire data and identify the overall semantic landscape of the data and maps.

4.3 Results

4.3.1 Theme and concepts

Leximancer was run to produce the most frequently occurring concepts from the texts and main themes of the journal articles as a cluster of concepts (Young, Wilkinson, & Smith, 2015). Leximancer produced 46 concepts and 11 themes from the 52 journal articles. The eight themes are marketing, consumers, journey, groups, purchase, price, strategies, companies, brand, and fat (see Table 4.2). We excluded common function words (e.g., *not* and *and*) and removed generic terms (e.g., *research*, *approach*, *results*, *examined*, *paper*, *suggest*, and *shows*), as these are commonly used in abstracts, without adding value to the concepts map.

Table 4.2 Themes and concepts in consumer food experience research.

Theme	Hits	Connectivity	Concepts
Food	4652	18,327	Food, health, public, obesity, unhealthy, eating, and nutrition
Marketing	3374	15,936	Marketing, children, advertising, industry, influence, and television
Consumers	2737	10,969	Consumers, consumption, social, choice, environment, and functional
Journey	597	2757	Journey and customer
Groups	338	1840	Groups
Purchase	612	1786	Purchase
Price	324	1708	Price
Strategies	273	1690	Strategies
Companies	220	1681	Companies
Brand	369	1373	Brand
Fat	286	1267	Fat

Moreover, we also merged most singular and plural words, such as *value/values* and *use/uses*, as these concepts are close in the semantic space. Importantly the most prominent concepts are themes linked in a group of concepts. Each theme consists of number of hits and connectivity, where hits indicate the number of text excerpts related to concepts, and connectivity shows the relative importance of each theme (Fisk, Fitzgerald, & Cokley, 2014; Mirzaei, Siuki, Gray, & Johnson, 2016).

After examining and evaluating the generated concepts and themes, we analyzed the conceptual map for a comprehensive view of the relationship among the main themes (Hair, Black, Babin, & Anderson, 2014) that the articles emphasized, namely, food, consumer, and marketing. The food theme is linked to concepts such as health, public, obesity, unhealthy, eating, and nutrition (e.g., Ludwig & Nestle, 2008). The consumer theme is linked to concepts such as consumption, social, choice, environment, and functional (e.g., Johnston, 2008; Vecchio & Annunziata, 2012). The marketing theme is linked to themes such as children, advertising, industry, influence, and television (e.g., Harris, Schwartz, & Brownell, 2010). The connectivity score indicates the relative importance of the themes (the most important is the top theme at 100%). In our analysis, the most prominent theme is *food*, with the most important concept related to healthy issues, since it is linked to *public*, *obesity*, *unhealthy*, *eating*, and *nutrition*.

4.3.2 Concepts linked to consumer food experience research

The Leximancer analysis shows that the three main themes (food, marketing, and consumers) generated numerous concepts. Some studies investigated the consumer food experience (Dobrenova, Grabner-Kräuter, & Terlutter, 2015; Igumbor et al., 2012; Larkin & Martin, 2016), although not explicitly emerging as a theme. Nevertheless, the outcome of the semantic analysis shows that the consumer food experience discussion encompasses food, consumers, marketing, and health (Table 4.3). The food concept was counted 3630 times (i.e., number of times mentioned in research linked to other concepts), with a 100% likelihood (i.e., referring to events that have a reasonable probability of occurring), thus the highest values of occurrence and likelihood, followed by consumers and marketing.

4.4 Discussion

Fig. 4.2 shows the key drivers of the unintended effects of enhancing the consumer experience. We investigate the themes in accordance with the ranking order and the link between the themes and concepts shown in Table 4.2.

4.4.1 Food

As Fig. 4.2 shows, the most prominent theme is food with links to health, public, obesity, unhealthy, eating, and nutrition mainly related to food marketing strategies

Table 4.3 Most frequent concepts.

Concept	Count	Likelihood (%)	Concept	Count	Likelihood (%)
Food	3630	100	Strategies	273	8
Consumers	1608	44	Nutrition	272	7
Marketing	1172	32	Eating	269	7
Health	1125	31	Functional	268	7
Children	808	22	Knowledge	267	7
Purchase	612	17	Weight	261	7
Advertising	610	17	Potential	252	7
Journey	579	16	Time	244	7
Consumption	524	14	Countries	238	7
Customer	474	13	Quality	234	6
Industry	469	13	Value	233	6
Obesity	461	13	Service	222	6
Public	453	12	Companies	220	6
Unhealthy	413	11	Change	216	6
Social	413	11	Policy	203	6
Choice	393	11	Role	200	6
Influence	392	11	Individual	197	5
Brand	369	10	Model	196	5
Environment	353	10	Development	178	5
Groups	338	9	School	173	5
Price	324	9	Government	146	4
Fat	286	8	Global	138	4
Television	275	8	Interest	129	4

that are often accused of contributing to the obesity epidemic (Swinburn et al., 2011). Food marketing strategies have encouraged consumers to eat more, enlarge their portions, and frequently consume ultraprocessed food (sweets, soft drinks, and snacks) (Ludwig & Nestle, 2008).

4.4.2 Marketing

The main theme of "marketing" relates to children, advertising, industry, influence, and television (Table 4.2). This aspect underlines that enhancing the consumer experience is based on disseminating information, increasing awareness of food with the highest nutritional and hedonic qualities (Marty et al., 2017).

Integrated marketing communications also focus on the specific category of children. This aspect reveals the influence on child behavior through an inventory method of advertising approaches including mobile advertising, social media marketing, branded websites, advertising, product placement, embedded advertising, event sponsorship, and television advertising (Harris et al., 2010).

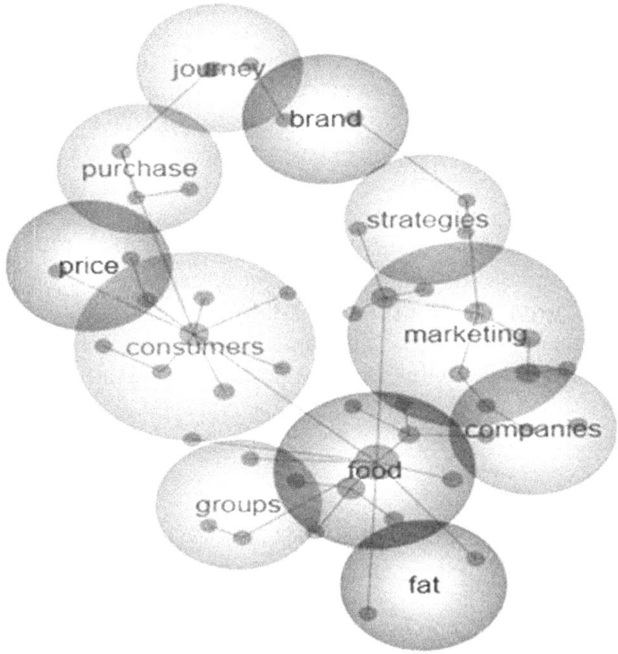

Figure 4.2 Thematic map of consumer food experience research in period 2000–2019.

4.4.3 Consumers

The next prominent theme is consumers linked to consumption, social, choice, environmental, and functional. This theme reflects the environmental and social issues, since customers are more sensitive to sustainable values influencing the overall customer experience. The trend is consistent with an overarching food system, including nutrition priorities for a healthy aging population, environmental sustainability, circular economy, and empowering local communities and minority groups. To note in this perspective is that citizens–consumers can express their social and environmental concerns while sustaining a commitment to a consumer ideology based on perpetual economic growth and sovereign consumer choice with few limitations (Johnston, 2008).

With a focus on functional food, their market launch is based on containing added technologically developed ingredients with specific health benefits, providing new innovative business opportunities (Niva, 2007). Vecchio and Annunziata (2012) underline that customer groups significantly differ in their evaluation of the perceived healthiness of functional foods and the importance attached to price and brand with implications on nutrition and health-related claims. This result reflects a critical evaluation in terms of the unintended consequences of improving the customer experience.

4.4.4 Customer journey

The customer journey is a customer-centered process that helps companies to not only improve their offering but also guide innovation, their allocation of resources, and business model transitions (Norton & Pine, 2013). Mapping the customer journey is instrumental to improving the customer experience at each point. Rosenbaum, Otalora, and Ramírez (2017) demonstrate that the customer journey helps a shopping mall's strategic management team to understand the more or less critical touchpoints for the customer experience.

Food is the primary variable that influences customers' memory and subsequently their behavior (Pantelidis, 2010). However, Han and Ryu's (2009) hierarchical multiple regression analysis of interactions shows that the quality of the food service and the physical environment are positive and significant determinants of restaurant satisfaction.

4.4.5 Groups

To analyze the marketing strategy concept, we consider the specific nature of the market segments related to the main theme "groups." In evidence, the food marketing industry specifically addresses children, although there is no clear and direct link with child obesity (Livingstone & Helsper, 2006).

Food marketing strategies often target specific ethnic groups and may produce changes in belief systems, such as the desirability of foods high in calories and low in nutrient density (Caprio et al., 2008).

4.4.6 Purchase

The purchase concept relates to the stages in the consumer marketing decision process. Berry, Mukherjee, Burton, and Howlett (2015) reveal the indirect effect of country of origin on purchasing intention through perceived (1) safety, (2) taste, and (3) freshness, which are positively associated, but the mediation effect reduces when consumers are presented with information on the similarity of products. The authors show the direct and indirect effects of the disclosure of objective information about the meat processing systems of competing countries (Berry et al., 2015).

4.4.7 Price

In analyzing the theme "price," most food products are not commodities and qualify as a branded product for the identified market segments. In the retail industry, lower food price leads to increased consumption (Ni Mhurchu, Blakely, Jiang, Eyles, & Rodgers, 2009), a policy observed in discount retailers. A substantial price reduction in cafeterias significantly increases the consumption of unhealthy food over time, and a substantial price increase reduces soft drink sales (Block, Chandra, Mcmanus, & Willett, 2010). The power of pricing also influences consumption through temporary price reductions and may increase calorie intake, since temporary sales promotions can induce an increase in the level of consumption (Chan, Narasimhan, & Zhang, 2008). A further incentive is the offer of quantity discounts with larger packaging sizes or multiunit packs (Vermeer, Steenhuis, & Seidell, 2009). Quantity discounts increase consumption, especially in the case of overweight consumers (Chandon & Wansink, 2002). Studies have found that people accelerate consumption of products perceived to have been purchased at a lower price (see, e.g., Thomas, Desai, and Seenivasan, 2010).

4.4.8 Strategies

In terms of strategies, we underline the role of marketing communications in influencing consumer decision-making based on product features compared with consumer well-being (Raghunathan, Naylor, & Hoyer, 2006). Of note is the importance of the authority of celebrity chefs in providing dietary and nutrition advice, while television programs promote an excessive amount of unhealthy fat, sugar, and salt (Phillipov, 2012). Relatedly the study also underline the effect of Master Chef Australia on increasing sales in supermarkets compared with the well-publicized nutritional guidelines from the main Australian health organizations recommending the reduction, removal, or substitution of butter from people's diet.

4.4.9 Companies

Another key aspect is related to "companies": retailers, restaurants, and food service companies can influence the ambient characteristics of the point of purchase and consumption (temperature, lighting, aroma, and noise) with marketing approaches promoting high fat and high calorie packaged food with minimal nutritional value (Weaver et al., 2014).

4.4.10 Brand

Branding has a substantial effect on the food product in relation to expected taste and health perceptions, beyond the description of nutritional information (Hoegg & Alba, 2006; Irmak, Vallen, & Robinson, 2011; Oakes, 2005). While nutrient composition and ingredients strongly influence health and taste expectations (Oakes, 2006; Oakes & Slotterback, 2005), most consumers do not access the nutrition information available (Grunert, Bolton, & Raats, 2012), which therefore does not have a substantial effect on food intake. Many health-based information sources are confusing or misunderstood by consumers (Levin & Gaeth, 1988; Wertenbroch, 1998). For example, the scientific approaches adopted influence the perceived effect of nutrient and ingredient composition (Mariotti, Kalonji, Huneau, & Margaritis, 2010). Consumers prefer simple front-of-packaging health information and guidelines, which may lead to unwanted health halos compared with more complete information (Ramanathan & Williams, 2007; Wansink & Chandon, 2006).

Health halos act on one positive aspect of the food item, leading customers to consider the food item as healthy, underestimating the calories, and thus overeating

(Andrews, Burton, & Kees, 2011). People expect they can eat more when the marketing, nutrition, and health information lead them to believe the target food is healthy. Food chain stores play an additional role in comparing brand values built through advertising campaigns: for example, a sandwich from Subway was perceived to have fewer calorie than McDonald's (Tangari, Burton, Howlett, Cho, & Thyroff, 2010).

Consumers, especially dieters, expect the combination of healthy and unhealthy food to contain fewer calories than unhealthy food alone (Chernev, 2010, 2011; Chernev & Gal, 2010).

4.4.11 Fat

A further key aspect is food composition: adding ingredients reduces the perception that the food is natural (Rozin, Fischler, & Shields-Argelès, 2009), but adding fat, sugar, and salt improves palatability (Stubbs & Whybrow, 2004). Food marketers have increased the supply of food rich in fat or added sugar, which has accounted for a large proportion of the added supply of calories in recent decades (Putnam, Allshouse, & Kantor, 2002).

4.4.12 Unintended consequences

The cross-comparison of findings reveals the unintended consequences of enhancing the consumer experience, since the key drivers identified affect consumer well-being in an unanticipated negative way. More specifically one of the emerging unintended consequences of the marketing practices determined is the obesity epidemic, with significant socioeconomic effects. In analyzing the prominent theme of consumer, we revealed environmental issues concerning the overarching food system but also underlining that the consumer experience does not always focus on an ideal vision based on value, community, inclusivity, and social justice. Thus the enhancement of the customer experience is frequently based on a materialistic orientation.

Customers' unaware decision processes result from limited health-based information on food composition, as indicated in the discussion concerning the themes of "brand" and "fat." In analyzing the theme related to "strategies," a further determinant emerged, namely, the presence of influencers and authorities alters consumer decision-making processes. Finally in analyzing "price," consumer behavior is influenced by price reductions and incentive systems that may lead to overeating.

4.5 Conclusions

This chapter has aimed to investigate the key drivers of the unintended consequences of enhancing the food customer experience through a text-mining procedure for analyzing and visualizing concepts and themes. To identify the factors that influence such unintended consequences, the conceptual map highlights the links with key marketing variables (such as brand, price, and customer experience), underlining the effects of health-based information, price reduction, and incentive systems, and the role of authorities acting as influencers. The strategic orientation of companies in service design is a further key driver with implications on consumer behavior.

We also reflect on the marketing practices aimed at vulnerable segments (children and ethnic groups), and the presence of an inventory method of advertising to engage and involve children.

Through a cross-comparison of the key drivers, the unintended consequences of enhancing the consumer experience emerge, including obesity, materialistic orientation, customers' unaware decision processes, and overeating.

In terms of future research, an emerging theme is analyzing conscious and unconscious marketing tactics involving key players in the food industry and consumer associations, and evaluating the impact on consumer behavior. To formulate innovative dietary recommendation policies, further studies should investigate food choice mechanisms in a more holistic perspective based on a multidisciplinary approach through integrating behavioral studies with neurobiological data studies.

References

Andrews, J. C., Burton, S., & Kees, J. (2011). Is simpler always better? Consumer evaluations of front-of-package nutrition symbols. *Journal of Public Policy & Marketing*, *30*, 175–190. Available from https://doi.org/10.1509/jppm.30.2.175.

Angus, D., Rintel, S., & Wiles, J. (2013). Making sense of big text: A visual-first approach for analysing text data using Leximancer and Discursis. *International Journal of Social Research Methodology*, *16*, 261–267. Available from https://doi.org/10.1080/13645579.2013.774186.

Berry, C., Mukherjee, A., Burton, S., & Howlett, E. (2015). A COOL effect: The direct and indirect impact of country-of-origin disclosures on purchase intentions for retail food products. *Journal of Retailing*, *91*, 533–542. Available from https://doi.org/10.1016/j.jretai.2015.04.004.

Block, J. P., Chandra, A., Mcmanus, K. D., & Willett, W. C. (2010). Point-of-purchase price and education intervention to reduce consumption of sugary soft drinks. *American Journal of Public Health*, *100*, 1427–1433. Available from https://doi.org/10.2105/AJPH.2009.175687.

Buijzen, M., & Valkenburg, P. M. (2012). The intended and unintended effects of advertising on children. *The International Encyclopedia of Media Studies*. Available from https://doi.org/10.1002/9781444361506.wbiems132.

Campbell, C., Pitt, L. F., Parent, M., & Berthon, P. R. (2011). Understanding consumer conversations around ads in a Web 2.0 world. *Journal of Advertising*, *40*, 87–102. Available from https://doi.org/10.2753/JOA0091-3367400106.

Caprio, S., Daniels, S. R., Drewnowski, A., Kaufman, F. R., Palinkas, L. A., Rosenbloom, A. L., & Schwimmer, J. B. (2008). Influence of race, ethnicity, and culture on childhood obesity: implications for prevention and treatment: A consensus statement of Shaping

America's Health and the Obesity Society. *Diabetes Care*, *31*, 2211–2221. Available from https://doi.org/10.2337/dc08-9024.

Chan, T., Narasimhan, C., & Zhang, Q. (2008). Decomposing promotional effects with a dynamic structural model of flexible consumption. *Journal of Marketing Research*, *45*, 487–498. Available from https://doi.org/10.1509/jmkr.45.4.487.

Chandon, P., & Wansink, B. (2002). When are stockpiled products consumed faster? A convenience-salience framework of postpurchase consumption incidence and quantity. *Journal of Marketing Research*, *39*, 321–335. Available from https://doi.org/10.1509/jmkr.39.3.321.19111.

Chernev, A. (2010). Semantic anchoring in sequential evaluations of vices and virtues. *Journal of Consumer Research*, *37*, 761–774. Available from https://doi.org/10.1086/656731.

Chernev, A. (2011). The dieter's paradox. *Journal of Consumer Psychology*, *21*, 178–183. Available from https://doi.org/10.1016/j.jcps.2010.08.002.

Chernev, A., & Gal, D. (2010). Categorization effects in value judgments: Averaging bias in evaluating combinations of vices and virtues. *Journal of Marketing Research*, *47*, 738–747. Available from https://doi.org/10.1509/jmkr.47.4.738.

Currie, J., Dellavigna, S., Moretti, E., & Pathania, V. (2010). The effect of fast food restaurants on obesity and weight gain. *American Economic Journal: Economic Policy*, *2*, 32–63. Available from https://doi.org/10.3386/w14721.

Dobrenova, F. V., Grabner-Kräuter, S., & Terlutter, R. (2015). Country-of-origin (COO) effects in the promotion of functional ingredients and functional foods. *European Management Journal*, *33*, 314–321. Available from https://doi.org/10.1016/j.emj.2015.03.003.

Fisk, K., Fitzgerald, R., & Cokley, J. (2014). Controversial new sciences in the media: Content analysis of global reporting of nanotechnology during the last decade. *Media International Australia*, *150*, 156–166. Available from https://doi.org/10.1177/1329878X1415000127.

Fry, M.-L., & Polonsky, M. J. (2004). Examining the unintended consequences of marketing. *Journal of Business Research*, *57*, 1303–1306. Available from https://doi.org/10.1016/S0148-2963(03)00073-0.

Grunert, K. G., Bolton, L. E., & Raats, M. M. (2012). *Processing and acting on nutrition labeling on food. Transformative consumer research for personal and collective well-being* (pp. 333–351). New York: Routledge.

Hair, J. F., Black, W. C., Babin, B. J., & Anderson, R. E. (2014). *Multivariate data analysis*. Pearson.

Han, H., & Ryu, K. (2009). The roles of the physical environment, price perception, and customer satisfaction in determining customer loyalty in the restaurant industry. *Journal of Hospitality & Tourism Research*, *33*, 487–510. Available from https://doi.org/10.1177/1096348009344212.

Harris, J. L., Schwartz, M. B., & Brownell, K. D. (2010). Marketing foods to children and adolescents: licensed characters and other promotions on packaged foods in the supermarket. *Public Health Nutrition*, *13*, 409–417. Available from https://doi.org/10.1017/S1368980009991339.

Hoegg, J., & Alba, J. W. (2006). Taste perception: More than meets the tongue. *Journal of Consumer Research*, *33*, 490–498. Available from https://doi.org/10.1086/510222.

Igumbor, E. U., Sanders, D., Puoane, T. R., Tsolekile, L., Schwarz, C., Purdy, C., . . . Hawkes, C. (2012). "Big food," the consumer food environment, health, and the policy response in South Africa. *PLoS Medicine*, *9*, e1001253.

Irmak, C., Vallen, B., & Robinson, S. R. (2011). The impact of product name on dieters' and nondieters' food evaluations and consumption. *Journal of Consumer Research, 38*, 390–405. Available from https://doi.org/10.1086/660044.

Johnston, J. (2008). The citizen-consumer hybrid: Ideological tensions and the case of whole foods market. *Theory and Society, 37*, 229–270. Available from https://doi.org/10.1007/s11186-007-9058-5.

Kalache, A., & Kickbusch, I. (1997). A global strategy for healthy ageing. *World health, 50*, 4–5. <https://www.popline.org/node/270394>.

Kraak, V. I., Gootman, J. A., & McGinnis, J. M. (2006). *Food marketing to children and youth: Threat or opportunity?* National Academies Press. Available from < ttps://zodml.org/sites/default/files/Food_Marketing_to_Children_and Youth_Threat or_Opportunity.pdf > .

Larkin, D., & Martin, C. R. (2016). Caloric estimation of healthy and unhealthy foods in normal-weight, overweight and obese participants. *Eating Behaviors, 23*, 91–96. Available from https://doi.org/10.1016/j.eatbeh.2016.08.004.

Levin, I. P., & Gaeth, G. J. (1988). How consumers are affected by the framing of attribute information before and after consuming the product. *Journal of Consumer Research, 15*, 374–378. Available from https://doi.org/10.1086/209174.

Livingstone, S., & Helsper, E. J. (2006). Does advertising literacy mediate the effects of advertising on children? A critical examination of two linked research literatures in relation to obesity and food choice. *Journal of Communication, 56*, 560–584. Available from https://doi.org/10.1111/j.1460-2466.2006.00301.x.

Ludwig, D. S., & Nestle, M. (2008). Can the food industry play a constructive role in the obesity epidemic? *Jama, 300*, 1808–1811. Available from https://doi.org/10.1001/jama.300.15.1808.

Mariotti, F., Kalonji, E., Huneau, J. F., & Margaritis, I. (2010). Potential pitfalls of health claims from a public health nutrition perspective. *Nutrition Reviews, 68*, 624–638. Available from https://doi.org/10.1111/j.1753-4887.2010.00322.x.

Marty, L., Miguet, M., Bournez, M., Nicklaus, S., Chambaron, S., & Monnery-Patris, S. (2017). Do hedonic-versus nutrition-based attitudes toward food predict food choices? a cross-sectional study of 6-to 11-year-olds. *International Journal of Behavioral Nutrition and Physical Activity, 14*, 162. Available from https://doi.org/10.1186/s12966-017-0618-4.

Mirzaei, A., Siuki, E. H., Gray, D., & Johnson, L. W. (2016). Brand associations in the higher education sector: The difference between shared and owned associations. *Journal of Brand Management, 23*, 419–438. Available from https://doi.org/10.1057/bm.2016.14.

Monteiro, C. A., Moubarac, J. C., Cannon, G., Ng, S. W., & Popkin, B. (2013). Ultra-processed products are becoming dominant in the global food system. *Obesity Reviews, 14*, 21–28. Available from https://doi.org/10.1111/obr.12107.

Ni Mhurchu, C., Blakely, T., Jiang, Y., Eyles, H. C., & Rodgers, A. (2009). Effects of price discounts and tailored nutrition education on supermarket purchases: A randomized controlled trial. *The American Journal of Clinical Nutrition, 91*, 736–747. Available from https://doi.org/10.3945/ajcn.2009.28742.

Niva, M. (2007). 'All foods affect health': Understandings of functional foods and healthy eating among health-oriented Finns. *Appetite, 48*, 384–393. Available from https://doi.org/10.1016/j.appet.2006.10.006.

Norton, D. W., & Pine, B. J. (2013). Using the customer journey to road test and refine the business model. *Strategy & Leadership, 41*, 12–17. Available from https://doi.org/10.1108/10878571311318196.

Oakes, M. E. (2005). Stereotypical thinking about foods and perceived capacity to promote weight gain. *Appetite*, *44*, 317–324. Available from https://doi.org/10.1016/j.appet.2005.03.010.

Oakes, M. E. (2006). Filling yet fattening: Stereotypical beliefs about the weight gain potential and satiation of foods. *Appetite*, *46*, 224–233. Available from https://doi.org/10.1016/j.appet.2006.01.004

Oakes, M. E., & Slotterback, C. S. (2005). Too good to be true: Dose insensitivity and stereotypical thinking of foods' capacity to promote weight gain. *Food Quality and Preference*, *16*, 675–681. Available from https://doi.org/10.1016/j.foodqual.2004.09.001.

Pantelidis, I. S. (2010). Electronic meal experience: A content analysis of online restaurant comments. *Cornell Hospitality Quarterly*, *51*, 483–491. Available from https://doi.org/10.1177/1938965510378574.

Phillipov, M. (2012). Communicating health risks via the media: What can we learn from MasterChef Australia? *The Australasian Medical Journal*, *5*, 593. Available from https://doi.org/10.4066/AMJ.2012.1460.

Putnam, J., Allshouse, J., & Kantor, L. S. (2002). US per capita food supply trends: More calories, refined carbohydrates, and fats. *Food Review*, *25*, 2–15.

Raghunathan, R., Naylor, R. W., & Hoyer, W. D. (2006). The unhealthy = tasty intuition and its effects on taste inferences, enjoyment, and choice of food products. *Journal of Marketing*, *70*, 170–184. Available from https://doi.org/10.1509/jmkg.70.4.170.

Ramanathan, S., & Williams, P. (2007). Immediate and delayed emotional consequences of indulgence: The moderating influence of personality type on mixed emotions. *Journal of Consumer Research*, *34*, 212–223. Available from https://doi.org/10.1086/519149.

Rosenbaum, M. S., Otalora, M. L., & Ramírez, G. C. (2017). How to create a realistic customer journey map. *Business Horizons*, *60*, 143–150. Available from https://doi.org/10.1016/j.bushor.2016.09.010.

Rozin, P., Fischler, C., & Shields-Argelès, C. (2009). Additivity dominance: Additives are more potent and more often lexicalized across languages than are "subtractives". *Judgment and Decision Making*, *4*, 475. Available from <http://journal.sjdm.org/8113/jdm8113.html>.

Slow Food. 2019. Our history. Available from <https://www.slowfood.com/about-us/our-history/> Accessed 08.04.19.

Smith, A. E., & Humphreys, M. S. (2006). Evaluation of unsupervised semantic mapping of natural language with Leximancer concept mapping. *Behavior Research Methods*, *38*, 262–279.

Stubbs, R., & Whybrow, S. (2004). Energy density, diet composition and palatability: Influences on overall food energy intake in humans. *Physiology & Behavior*, *81*, 755–764. Available from https://doi.org/10.1016/j.physbeh.2004.04.027.

Suro, M. D. (1986, May 5). Romans protest McDonalds. *New York Times*.

Swinburn, B. A., Sacks, G., Hall, K. D., Mcpherson, K., Finegood, D. T., Moodie, M. L., & Gortmaker, S. L. (2011). The global obesity pandemic: Shaped by global drivers and local environments. *The Lancet*, *378*, 804–814. Available from https://doi.org/10.1016/S0140-6736(11)60813-1.

Tangari, A. H., Burton, S., Howlett, E., Cho, Y. N., & Thyroff, A. (2010). Weighing in on fast food consumption: The effects of meal and calorie disclosures on consumer fast food evaluations. *Journal of Consumer Affairs*, *44*, 431–462. Available from https://doi.org/10.1111/j.1745-6606.2010.01177.x.

Thomas, M., Desai, K. K., & Seenivasan, S. (2010). How credit card payments increase unhealthy food purchases: Visceral regulation of vices. *Journal of Consumer Research, 38*, 126–139. Available from https://doi.org/10.1086/657331.

Vecchio, R., & Annunziata, A. (2012). Italian consumer awareness of layer hens' welfare standards: A cluster analysis. *International Journal of Consumer Studies, 36*, 647–655. Available from https://doi.org/10.1111/j.1470-6431.2011.01040.x.

Vermeer, W. M., Steenhuis, I. H., & Seidell, J. C. (2009). Portion size: A qualitative study of consumers' attitudes toward point-of-purchase interventions aimed at portion size. *Health Education Research, 25*, 109–120. Available from https://doi.org/10.1093/her/cyp051.

Wansink, B., & Chandon, P. (2006). Can "low-fat" nutrition labels lead to obesity? *Journal of Marketing Research, 43*, 605–617. Available from https://doi.org/10.1509/jmkr.43.4.605.

Watson, M., Smith, A., & Watter, S. (2005). *Leximancer concept mapping of patient case studies. International conference on knowledge-based and intelligent information and engineering systems* (pp. 1232–1238). Springer.

Weaver, C. M., Dwyer, J., Fulgoni Iii, V. L., King, J. C., Leveille, G. A., Macdonald, R. S., ... Schnakenberg, D. (2014). Processed foods: Contributions to nutrition. *The American Journal of Clinical Nutrition, 99*, 1525–1542. Available from https://doi.org/10.3945/ajcn.114.089284.

Wertenbroch, K. (1998). Consumption self-control by rationing purchase quantities of virtue and vice. *Marketing Science, 17*, 317–337. Available from https://doi.org/10.1287/mksc.17.4.317.

Young, L., Wilkinson, I., & Smith, A. (2015). A scientometric analysis of publications in the Journal of Business-to-Business Marketing 1993–2014. *Journal of Business-to-Business Marketing, 22*, 111–123. Available from https://doi.org/10.1080/1051712X.2015.1021591.

Crafting food shopping experiences: the case of food halls in luxury departments stores

5

Eirini Koronaki and Prokopis K. Theodoridis
Department of Business Administration of Food and Agricultural Enterprises, University of Patras, Patras, Greece

5.1 Introduction

Consumers' demand for unique and immersive experiences is constantly increasing. And while initially this referred mostly to high-involvement product categories and their brands, it has now expanded and is relevant for the majority of consumption moments individuals experience in their everyday life. These unique experiences are crafted and provided through the satisfaction of either hedonic or utilitarian dimensions, or even a combination of the above (Varshneya, Das, & Khare, 2017).

Brands are thus reexamining the way they connect to their customers and the moments of connection they are providing them with. This connection can refer to the brand's identity on the one hand and to the context in which the brand is being purchased on the other. In fact, the design and creation of unique selling outlets for brands has become a crucial parameter to their success in some cases (Bagdare & Jain, 2013; Bustamante & Rubio, 2017; Jain & Bagdare, 2009; Mathwick, Malhotra, & Rigdon, 2002). An increasing number of brands are investigating how to construct their outlets in a way to provide the desired experiential elements and to create strong and impactful experiences for their customers (e.g., Jones, Comfort, Clarke-Hill, & Hillier, 2010; Naylor, Kleiser, Baker, & Yorkston, 2008; Srinivasan & Srivastava, 2010).

Interestingly this urge for experiences has also affected the food industry. And while initially the purchase of food products, which are nondurable, are frequently purchased and are in most cases linked to decreased degrees of interest and involvement, would be classified as a task-oriented purchase (Machleit & Eroglu, 2000), this is something that has started to change. An increasing number of consumers no longer see shopping for food as the low involvement, repeat purchase, and price-driven decision making (Van Camp, Hooker, & Souza-Monteiro, 2010) that it was. This does not solely refer to the small segment of the so-called "foodies," defined as an individual who links food to culture and social status (Yeoman & McMahon-Beatte, 2016), considers food able to enhance food and sensations (Yeoman, 2012), and who significantly cares about the experiential dimensions (Lang, Stanton,

Case Studies on Food Experiences in Marketing, Retail, and Events. DOI: https://doi.org/10.1016/B978-0-12-817792-1.00005-8

Lancioni, & Herbst, 2010). This tendency can now also be identified in an increasing number of mainstream food consumers.

Food products have an innate connection to the senses, consumers are increasingly intrigued by the inclusion of experiential dimensions in the respective food retail outlets, from supermarkets, artisanal stores, and hawkers to open-air markets. Some indicative examples of experiential elements in food retail outlets include artisanal goods, connection with local restaurants and menus which are recommended by renowned chefs (Business Insider, 2016). Despite the fact that food retail outlets have started to build and invest in this direction, there are still many activities that can be done for the creation of unique immersive food retail outlet experiences.

A brand category which has been established as a master of designing unique customer experiences is the luxury one. Especially in the context of the retail experience, luxury brands are able to include a variety of elements, to please their customers' senses and to effectively represent the brand's universe. Through features of art and magic (Dion & Arnould, 2011), such outlets are are to foster hedonic and aesthetic dimensions of shopping (Kim, Park, Lee, & Choi, 2016). Based on the notions of exclusivity and uniqueness, they try to give the customer a solid reason for their visit (Nobbs, Moore, & Sheridan, 2012). One of the luxury categories which have been constantly gaining ground and attention lately are the food halls in luxury department stores, offering both a variety of restaurants, cafes, and bars and a wide range of products. These food halls are a very good and indicative example of an immersive food retail outlet experience and can function as a guiding case for other types of outlets as well. This is because they are enhancing the experiential dimension of shopping by offering "emotional, aesthetic, hedonic, psychological, and pleasurable benefits that exist beyond utilitarian, rational, and economic benefits" (Lang & Hooker, 2013, p. 642). Shopping in stores that provide such strong emotional experiences are in most cases remembered by consumers and have been linked to various benefits regarding positive attitudes and behaviors (Pantano & Gandini, 2018).

Throughout this chapter, the ways to create such aesthetically pleasing universes in the food halls of luxury department stores are examined. First, some basic concepts on food shopping are presented. Then, the chapter analyzes how food retail outlets can contribute to the creation of pleasing food shopping experiences through their basic elements and how this can be combined with the four types of experiences. Some representative examples of food halls of luxury department stores are presented and discussed. Finally the basic benefits of such food retail outlets are presented.

5.2 Heading toward experiential food shopping

Traditionally food shopping was perceived as a type of purchase of low involvement and interest, due to its nondurable nature and the frequency with which it was bought (Holbrook & Hirschman, 1982). This context of low involvement is highly linked to the utilitarian dimensions of consumption (Babin, Darden, & Griffin, 1994). This means that consumers in such a case would mostly care about performing the task and are mostly affected by dimensions such as the product's price, its quality, the existing variety, and the location of the store. As a result, the viewpoint prevailed that the outlets in which such purchases are made would be of a similar nature. For example, it has been noted that grocery and supermarket stores are not as exciting as entertaining outlets (Machleit & Eroglu, 2000).

However, this is something that has started to change. It has been identified that building solely on functional and rational elements is not sufficient to create satisfactory outcomes in consumers' minds (Varshneya & Das, 2017). It is a fact that when consumers shop they do not only seek out for the functional benefits, but they are also interested in the hedonic ones (Babin et al., 1994; Varshneya et al., 2017). The hedonic elements of consumption have been gaining increased attention (Voss, Spangenberg, & Grohmann, 2003). This describes the desire for experience (Tynan, McKechnie, & Chhuon, 2010), through notions such as "enjoyment, excitement, captivation, escapism, and spontaneity" (Babin et al., 1994).

Enhancing those hedonic elements has been linked to increased levels of perceived satisfaction (Chitturi, Raghunathan, & Mahajan, 2008). It has been noted that due to the increased competition in the sector and the fact that many retailers compete on price, it has become crucial to understand what satisfies the customer The customer's satisfaction has been linked both to the consumers' expectations and to the products' characteristics (Lang & Hooker, 2013). At this point it is interesting to mention a difference noted in the literature between a customer who is merely satisfied and one who is completely satisfied (Jones & Sasser, 1995) in terms of the positive outcomes that might emerge. These outcomes emerge as a result of what the consumer thinks, feels, and responds when shopping (Solomon, 2009, p. 247). They might include various benefits such as word-of-mouth promotion and increasing repurchase intentions and subsequent levels of loyalty (Chitturi et al., 2008), retail preferences and choice of store, shopping satisfaction, willingness to pay and spending (Machleit & Mantel, 2001), and profit (Ailawadi, Beauchamp, Donthu, Gauri, & Shankar, 2009). Moreover, the consumer's self-image can be enhanced through such a visit parallel to positive social interactions.

While this is something that has been long established for nonfood retail outlets, it is only recently that this dimension has been identified as crucial in the food retail sector, making many researchers and managers eager to identify ways to create various emotion types while shopping. Initially the experiential element in food shopping was only considered relevant in the case of the specialty segment food retailing (Lang et al., 2010). The demand for experiential elements would depend on the retail setting, the product category, the customer's type and the level of involvement. However, the viewpoint prevailed that this is relevant not just for customers of specialty stores but for any type of food retailer (Lang & Hooker, 2013).

Food retail outlets can thus leverage on the importance placed on hedonic elements of consumption and also gain from the benefits they are going to have in terms of consumer outcomes. Building on that, the design of the retail outlet, the combinations of products and the way the service is provided aims to provide unique experiences to the customer. This does not necessarily mean that all elements of experiential dimensions need to be included. But if the dimensions of the food retail outlet's brand identity and its audience's image about it are understood and clear, one will be able to invest in the right type of experiences. In an era of increased competition for food retailers, both on an electronic format and on specialty stores increasing, being able to take advantage of this opportunity and keep one's existing audience can be a significant benefit not only in terms of improved performance and also in some cases a matter of survival.

5.3 Elements for the design of retail outlets: including the experiential dimension

The question of how to create a food retail outlet providing both hedonic and functional benefits emerges. Before going into the inclusion of experiential elements in a food retail outlet, the elements on which a retail outlet is designed are presented. There are three dimensions to consider for the design of a retail outlet (Eroglu, Machleit, & Davis, 2001; Fiore & Kim, 2007) (see Fig. 5.1). The first one refers to atmospherics and includes ambient and nonstructural elements. Those less tangible

Atmospheric elements	Physical elements	Social elements
• Ambient and nonstuctural elements •The decoration •The light •The music • Any scents used	• Physical space and design •The outlet's layout •The product assortment • How the products are displayed • How the product is acquired by the customer	• Human interactions • Interaction with employees • Interaction with other customers

Figure 5.1 Elements of retail outlets.
Source: Based on Machleit, K. A., & Mantel, S. P. (2001). Emotional response and shopping satisfaction: Moderating effects of shopper attributions. *Journal of Business Research*, *54*(2), 97–106. <https://doi.org/10.1016/S0148-2963(99)00093-4>; Lang, M., Stanton, J., Lancioni, R., & Herbst, K. (2010). The role of hedonistic and emotional response in food shopping. In *Conference proceedings of the 17th recent advances in retailing and services sciences conference*, Istanbul, Turkey, July 2–5.

aspects are able to create a desired mood and style through the decoration, the light, the music, and any scents used (Machleit & Mantel, 2001; Parsons, Ballantine, Ballantine, Jack, & Parsons, 2010). The second one refers physical elements and the design used in the space for the creation of a strong experience. This includes issues such as the outlet's layout, the product assortment, how the products are displayed (merchandising), and how the product is acquired by the customer (Machleit & Mantel, 2001). The third one refers to the social elements including any human interactions that might be fostered and cultivated, both with employees and with other customers (Lang et al., 2010).

Since the shopping experience refers to the intersection between the retail environment and consumer behavior (Lang & Hooker, 2013), the retail outlets which aim to provide an extraordinary experience need to include such elements in the three aforementioned dimensions (Sachdeva & Goel, 2015). This is based on the fact that the overall consumption experience is based on more than just rational elements and is combined with multisensory and emotional aspects of the experience (Holbrook & Hirschman, 1982). Some of dimensions included are adventure, joy, the satisfaction of various roles and social elements (Arnold & Reynolds, 2003; Yakhlef, 2015). It has been stated that experiential retail outlets are not just trying to sell products, but they are trying to tell a story through the additional elements they include in their environment (Jahn, Nierobisch, Toporowski, & Dannewald, 2018). Some of those elements might be cues to the company's heritage, services that are customized, and exclusive product lines (Nierobisch, Toporowski, Dannewald, & Jahn, 2017).

To understand how such experiences can be designed and delivered within an outlet, some basic concepts of the experience literature need to be integrated. A significant concept in this literature relates to the various types of experiences that a company can design, classified as entertaining, educational, escapist, or aesthetic (Pine & Gilmore, 1998). This classification was made depending on the degree of customer participation and connection they can provide, referring to active or passive participation and absorption or immersion, respectively. Entertaining experiences are the ones characterized by a passive participation and absorption, such as watching TV or attending a concert (Pine & Gilmore, 1998). Educational experiences are the ones with active participation and absorption, such as attending a class or taking a lesson (Pine & Gilmore, 1998). Escapist refer to active participation and immersion, such as acting in a play, playing in an orchestra or descending the Grand Canyon, or the famous Samaria Gorge in south-western Crete (Pine & Gilmore, 1998). Finally aesthetic ones refer to passive participation and immersion, such as a visitor to an art gallery or a tourist who merely sees the Grand Canyon from where he is standing (Pine & Gilmore, 1998, p. 102).

5.4 Methodology

The present chapter thus seeks to answer the following research question: how can immersive and unique food shopping experienced be designed in a retail setting?

To answer the research question, a multiple-case study design was adopted, to acknowledge the complexity of the phenomenon (McCutcheon & Meredith, 1993), and to identify emerging patterns (Eisenhardt & Graebner, 2007). For the multiple-case study, the top five luxury department stores included in the Sybarite Report (The Sybarite Report, 2019) (Harrods in London, United Kingdom, Le Bon Marche in Paris, France, Liberty In London, United Kingdom, Skp in Hong Kong, China, and Selfridges in London, United Kingdom), which identifies the most successful luxury department stores globally, were selected. Data sources were secondary data such as their websites, Facebook pages, and industry articles, to get a deeper understanding of the cases (Patton, 2002). The data were analyzed in an attempt to identify the common behaviors among department stores regarding the four types of experiences in the three basic retail elements, identified in the literature. For the analysis on the experiential elements, multiple sources of information we used. As recommended by Yin (2003, p. 137) each case was studied and analyzed before performing cross-case analysis to detect patterns and then were proceeded with cross-case analysis. Data gathered from these sources were used to understand the cases and this analysis was driven by the research questions. The data was grouped, recorded, and analyzed according to the three experiential retail dimensions. This improved the possibility of reaching accurate and reliable theory (Eisenhardt, 1989).

In the following section it is explored how food halls of luxury department stores apply experiential elements in their settings.

5.5 Applying the framework in luxury department stores food halls

5.5.1 Analysis of atmospheric elements

The atmospheric elements have been traditionally leveraged on by luxury department food halls. An indicative example is the Selfridges food hall, which is characterized with extravagance and is increasingly using the scents of the hot food or traditional desserts they prepare as a means to create a unique experience. Similarly in the Mini British Food Hall in Liberty, the fresh British products create multisensory experiences.

The decoration is another significant example frequently used for the creation of unique experiences. Indicatively La Grande Épicerie, which is the food hall of the Parisian luxury department store Le Bon Marché has decorated the place in a way to please the senses and to provide a pleasing experience (Business Insider, 2016). Similarly Harrods' food hall in London includes ornate display cases and painted ceilings. In terms of the light element, another indicative example is the area Rendez-vous of the department store SKP in Beijing, named Le Printemps du Gout, which is located on the fourth floor offering a beautiful view of the city (The Sybarite Report, 2019).

Numerous experiences can be organized in the physical dimensions area described above. Indicatively the cases in which the consumer can walk around in a beautiful space with exquisite decoration, and filled with light can be seen as an entertaining experience. If the outlet chooses to use music of a specific country to combine with the food section, this can also be seen as an educational experience, in which the consumer would learn about the music of a region. For example, this can be identified in Mini British Food Hall's brand identity elements in the department store Liberty (Liberty London, 2019), where people can learn about the British culture, geography, and music through the elements of the retail store.

5.5.2 Analysis of physical elements

Physical elements are extensively used in food halls. An indicative example in terms of product assortment of the Rendez-vous area in the department store SKP with its wine bar and cellar and artisanal cheese room with cheese-making facilities. Another example regarding the product assortment is the department store Harrods' food hall. Even though they specialize in luxury food, they have positioned themselves as a choice for tourists to get a souvenir, offering numerous restaurants, products, spirits, and drinks from local producers as well (Business Insider, 2016). Another interesting element in the assortment section relates with having international products to offer. For example, in the food hall La Grande Epicerie, of the luxury department store Le Bon Marche in Paris, a variety of international specialty foods are offered such as Spanish ham, British shortbread, Italian tomato sauce, and American Marshmallow (Frommers, 2019). The department store Harrods is a very good example for product display as well, since it has been known for the eye-pleasing way the items are displayed in the food hall (Frommers, 2019).

As far as the outlet's layout is concerned, there are steps taken in that direction as well. For example, the department store Liberty, has installed impressive arrangements in their food hall, which people can visit as a break from shopping (The Sybarite Report, 2019).

A variety of experiences can be designed through the physical elements as well. The variety of products from all around the world can be seen as an entertaining experience, increased through a pleasing store layout, display and also acquisition. Through the layout and the assortment the visitor can engage with educational experiences, especially if this is combined with a class or a set of lessons offered by the food hall. The aesthetic experience can also be enhanced through the display of the products. The escapist experience can also be satisfied through associations created within the food hall. For example, the department store Harrods had created an event in which fashion and food were creatively combined and linked to various events organized in the store and also with collaborating partners (Luxury Daily, 2012). Participating in such an event can function as an escapist experience for the consumer.

5.5.3 Analysis of social elements

Significant elements are also added in the social dimensions in such food halls. For example, in the fourth floor of Rendez-vous in the department store SKP, the visitor has the chance to eat or drink something and thus interact both with employees and with other customers (The Sybarite Report, 2019). Another point of interaction in SKP is also created with the inclusion of traditional food corners, in which the visitors can also interact with the artisan while shopping, such as the cheese corner in Rendez-vous. The degree of knowledge that the food halls' employees have is another significant point of interaction. Indicatively in the department store Selfridge's, the food hall's employees' knowledge on teas and coffees is acknowledged. The collaboration with acknowledged chefs and artisans in their fields is another significant tactic used by such food halls, with the indicative example of the Champagne & Oyster Bar by Caviar House & Prunier in Selfridges (Selfridges, 2019).

Numerous experiences can be delivered through social elements as well. For example through special events organized in these food halls, which consumers can visit with friends and positively interact with the brand can create entertaining experiences. Furthermore though the promotion of certain ideals within the food hall, such as the initiative organized by the department store Liberty, to support British producers and cooperate in brand identity and design issues can provide an educational experience for the consumer, both through the interaction with other employees and through the interaction with other customers. Furthermore if we consider the collaboration with star chefs, aesthetic experiences can be offered to the visitors when in the store, as is the case with Michelin chefs' collaboration with the department store Harrods (Frommers, 2019).

From the analysis it becomes evident that a variety of elements are included in food halls of luxury department stores, frequently also combined with various types of experiences. Thus visitors in such food halls can emerge into the shopping experience and enjoy the variety of benefits linked to that.

The elements to be included vary as presented above from atmospheric elements such as decoration, light, music and scents to physical elements such as the design, the outlet's layout, the variety of products, and how they are displayed and acquired, and also social elements based on the interactions both with other customers and with employees. Despite the fact that the examples used referred to the food halls of luxury department stores, they can function as guidance for any food retail outlet which aims to provide a unique experience to its customers and gain from the multitude of benefits that can be gained from providing the consumer with such immersive experiences. In Fig. 5.2, experiential elements which can be added to retail dimensions are presented, which arise after the examination and analysis of the selected cases and their connection to existing literature in the field.

5.6 Discussion and conclusions

There are numerous reasons for which brands should invest in the design of such experiential food retail outlets. Various benefits can be gained in case an outlet manages to create a holistic environment, in which consumers would gain both functional and hedonic benefits (Khan & Rahman, 2016; Pantano & Gandini,

Type of retail element	Atmospheric elements	Entertainment	Educational	Escapist	Aesthetic
Dimensions of retail element	Decoration Light Music Any scents used	• Hosting music concerts with music from specific regions linked to food types. • Use of multimedia to show pictures from those regions • Background music from those regions	• Lessons on how to combine decorative elements and art-de- table • Lessons on associating music and food	• Invitation to participate in a musical soiree "music & food" themed hosted in the store • Take part in setting up a decorative artistic element in the store (e.g.,installation) • Organize a trip to specific regions for smelling and tasting experiences	• Host a food and light-themed exhibition with art paintings within the store • Host an exhibition with paintings from a specific area within the store • Admire a decorative artistic element (painting, sculpture,etc.) • Light installation • Use of scents linked to a specific geographic area • Admire the view to a part of the city
Type of retail element	Physical elements	Entertainment	Educational	Escapist	Esthetic
Dimensions of retail element	Design Outlet's layout Product assortment Product display Product acquisition	• Videos about food display and assortment • Serving the food in an impressive and experiential manner	• Classes about food combinations and display • Advice and guidance on a one-to-one basis from expert employees	• Organize a "set the table" contest • Organize a "best assortment" contest • Organize trips to regions to buy food from the country of origin	• Art installation themed on food display and layout • Photo exhibition with a food display theme
Type of retail element	Social elements	Entertainment	Educational	Escapist	Esthetic
Dimensions of retail element	Any interaction with employees and with other customers	• Hosting parties with specific themes • Watch a chef prepare a dinner	• Organize "learn from others" sessions about recipes food combinations • Host classes from famous chefs • Host classes from expert employees on specific themes	• Organize trips to visit food markets • Organize trips and tours to chef's home country • Host dinners in various secluded and unique parts of the world	• Organize visits to exhibition within the store for specific groups of customers with similar tastes • Organize visits to exhibition within the store with expert employees

Figure 5.2 The types of experiences that can be organized in the three retail outlet elements.

2018). Both of those benefits have the ability to provoke both rational and emotional responses and thus increase the perceived value gained (Stein & Ramaseshan, 2016). The ways through which the consumer engages to the brand can have various levels in such a context, ranging from emotional, educational, physical, or intellectual types (Voss et al., 2003). Two important dimensions of engagement should be considered at this point are store engagement on the one hand and consumer engagement on the other. Customer engagement is positively linked to the servicescapes (Choudhary & Singh, 2017). Moreover, when consumers have a high degree to store engagement, this positively affects their customer engagement in the context of food retail stores. On the other hand, the quality of the food can positively affect the store engagement. It has in fact been identified that elements such as pleasure and arousal can have positive influences toward the retailer (Dawson, Bloch, & Ridgway, 1990).

To summarize, this chapter contributed to existing knowledge on various levels. First, the theory on experiential and luxury retails spaces was used as a framework of analysis in the food retail outlet context (Naylor et al., 2008; Srinivasan & Srivastava, 2010). The three basic elements of a retail space were combined with the four types of experiences a brand can provide (Lang et al., 2010; Machleit & Mantel, 2001). This understanding can be beneficial and leveraged on for premium food retail outlets, and also for mainstream ones, which aim to enhance their customer experience. Furthermore luxury, premium, and mainstream food retail outlets can get guidance from the identification of such practices, in their attempt to provide unique customer experiences. Consequently through an understanding of how a variety of experiences can be designed with the assistance of the store elements, each brand can choose which dimensions to build on and develop depending on both the story and identity of their brand, and also the value that their customers seek for in their visits in their outlets.

References

Ailawadi, K. L., Beauchamp, J. P., Donthu, N., Gauri, D. K., & Shankar, V. (2009). Communication and promotion decisions in retailing: A review and directions for future research. *Journal of Retailing*, *85*(1), 42–55. Available from https://doi.org/10.1016/j.jretai.2008.11.002.

Arnold, M. J., & Reynolds, K. E. (2003). Hedonic shopping motivations. *Journal of Retailing*, *79*(2), 77–95. Available from https://doi.org/10.1016/j.jbusres.2006.03.006.

Babin, B. J., Darden, W. R., & Griffin, M. (1994). Work and/or fun: Measuring hedonic and utilitarian shopping value. *Journal of Consumer Research*, *20*(4), 644–656. Available from https://doi.org/10.1086/209376.

Bagdare, S., & Jain, R. (2013). Measuring retail customer experience. *International Journal of Retail & Distribution Management*, *41*(10), 790–804. Available from https://doi.org/10.1108/IJRDM-08-2012-0084.

Business Insider. (2016). World's best-department store food halls. Retrieved from <https://www.businessinsider.com/worlds-best-department-store-food-halls-2016-9#lafayette-gourmet-at-galeries-lafayette-paris-france-9>.

Bustamante, J. C., & Rubio, N. (2017). Measuring customer experience in physical retail environments. *Journal of Service Management*, *28*(5), 884–913. Available from https://doi.org/10.1108/JOSM-06-2016-0142.

Chitturi, R., Raghunathan, R., & Mahajan, V. (2008). Delight by design: The role of hedonic versus utilitarian benefits. *Journal of Marketing*, *72*(3), 48–63. Available from https://doi.org/10.1509/JMKG.72.3.048.

Choudhary, S., & Singh, T. (2017). Role of retail store attributes on store and customer engagement in food industry. *International Journal of Marketing & Business Communication*, *6*(3).

Dawson, S., Bloch, P. H., & Ridgway, N. (1990). Shopping motives, emotional states, and retail outcomes. *Journal of Retailing*, *66*(4), 408–427.

Dion, D., & Arnould, E. (2011). Retail luxury strategy: Assembling charisma through art and magic. *Journal of Retailing*, *87*(4), 502–520. Available from https://doi.org/10.1016/j.jretai.2011.09.001.

Eisenhardt, K. M. (1989). Building theories from case study research. *The Academy of Management Review*, *14*(4), 532–550.

Eisenhardt, K. M., & Graebner, M. E. (2007). Theory building from cases: Opportunities and challenges. *Academy of Management Journal*, *50*(1), 25–32. Available from https://doi.org/10.5465/amj.2007.24160888.

Eroglu, S. A., Machleit, K. A., & Davis, L. M. (2001). Atmospheric qualities of online retailing: A conceptual model and implications. *Journal of Business research*, *54*(2), 177–184. Available from https://doi.org/10.1016/S0148-2963(99)00087-9.

Fiore, A. M., & Kim, J. (2007). An integrative framework capturing experiential and utilitarian shopping experience. *International Journal of Retail & Distribution Management*, *35*(6), 421–442. Available from https://doi.org/10.1108/09590550710750313.

Frommers. (2019). World's best department store food halls. Retrieved from <https://www.frommers.com/slideshows/822457-world-s-best-department-store-food-halls>.

Holbrook, M. B., & Hirschman, E. C. (1982). The experiential aspects of consumption: Consumer fantasies, feelings, and fun. *Journal of Consumer Research*, *9*(2), 132–140. Available from https://doi.org/10.1086/208906.

Jahn, S., Nierobisch, T., Toporowski, W., & Dannewald, T. (2018). Selling the extraordinary in experiential retail stores. *Journal of the Association for Consumer Research*, *3*(3), 412–424. Available from https://doi.org/10.1086/698330.

Jain, R., & Bagdare, S. (2009). Determinants of customer experience in new format retail stores. *Journal of Marketing & Communication*, *5*(2).

Jones, P., Comfort, D., Clarke-Hill, C., & Hillier, D. (2010). Retail experience stores: Experiencing the brand at first hand. *Marketing Intelligence & Planning*, *28*(3), 241–248. Available from https://doi.org/10.1108/02634501011041408.

Jones, T. O., & Sasser, W. E. (1995). Why satisfied customers defect. *Harvard Business Review*, *73*(6), 88.

Khan, I., & Rahman, Z. (2016). Retail brand experience: Scale development and validation. *Journal of Product & Brand Management*, *25*(5), 435–451. Available from https://doi.org/10.1108/JPBM-07-2015-0943.

Kim, S., Park, G., Lee, Y., & Choi, S. (2016). Customer emotions and their triggers in luxury retail: Understanding the effects of customer emotions before and after entering a luxury shop. *Journal of Business Research*, *69*(12), 5809–5818. Available from https://doi.org/10.1016/j.jbusres.2016.04.178.

Lang, M., & Hooker, N. H. (2013). An empirical test of experiential shopping in food retailing. *British Food Journal*, *115*(5), 639–652. Available from https://doi.org/10.1108/00070701311331553.

Lang, M., Stanton, J., Lancioni, R., & Herbst, K. (2010). The role of hedonistic and emotional response in food shopping. In Conference proceedings of the 17th recent advances in retailing and services sciences conference, Istanbul, Turkey, July 2–5.

Liberty London. (2019). Mini British food hall. Retrieved from <https://www.libertylondon.com/uk/instore-pages/mini-british-food-hall-meet-the-producers.html>.

Luxury Daily. (2012, October 2). Harrods marries fashion, food in digital campaign. *Luxury Daily*. Retrieved from <https://www.luxurydaily.com/harrods-marries-fashion-food-in-digital-campaign/>.

Machleit, K. A., & Eroglu, S. A. (2000). Describing and measuring emotional response to shopping experience. *Journal of Business Research*, *49*(2), 101–111. Available from https://doi.org/10.1016/S0148-2963(99)00007-7.

Machleit, K. A., & Mantel, S. P. (2001). Emotional response and shopping satisfaction: Moderating effects of shopper attributions. *Journal of Business Research*, *54*(2), 97–106. Available from https://doi.org/10.1016/S0148-2963(99)00093-4.

Mathwick, C., Malhotra, N. K., & Rigdon, E. (2002). The effect of dynamic retail experiences on experiential perceptions of value: An Internet and catalog comparison. *Journal of retailing*, *78*(1), 51–60. Available from https://doi.org/10.1016/S0022-4359(01)00066-5.

McCutcheon, D. M., & Meredith, J. R. (1993). Conducting case study research in operations management. *Journal of Operations Management*, *11*(3), 239–256. Available from https://doi.org/10.1016/0272-6963(93)90002-7.

Naylor, G., Kleiser, S. B., Baker, J., & Yorkston, E. (2008). Using transformational appeals to enhance the retail experience. *Journal of Retailing*, *84*(1), 49–57. Available from https://doi.org/10.1016/j.jretai.2008.01.001.

Nierobisch, T., Toporowski, W., Dannewald, T., & Jahn, S. (2017). Flagship stores for FMCG national brands: Do they improve brand cognitions and create favorable consumer reactions? *Journal of Retailing and Consumer Services*, *34*, 117–137. Available from https://doi.org/10.1016/j.jretconser.2016.09.014.

Nobbs, K., Moore, C. M., & Sheridan, M. (2012). The flagship format within the luxury fashion market. *International Journal of Retail & Distribution Management*, *40*(12), 920–934. Available from https://doi.org/10.1108/09590551211274928.

Pantano, E., & Gandini, A. (2018). Shopping as a "networked experience": An emerging framework in the retail industry. *International Journal of Retail & Distribution Management*, *46*(7), 690–704. Available from https://doi.org/10.1108/IJRDM-01-2018-0024.

Parsons, A. G., Ballantine, P. W., Ballantine, P. W., Jack, R., & Parsons, A. G. (2010). Atmospheric cues and their effect on the hedonic retail experience. *International Journal of Retail & Distribution Management*. Available from https://doi.org/10.1108/09590551011057453.

Patton, M. Q. (2002). *Qualitative research and evaluation methods*. Thousand Oakes, CA: Sage.

Pine, B. J., & Gilmore, J. H. (1998). Welcome to the experience economy. *Harvard Business Review*, *76*, 97–105.

Sachdeva, I., & Goel, S. (2015). Retail store environment and customer experience: A paradigm. *Journal of Fashion Marketing and Management*, *19*(3), 290–298. Available from https://doi.org/10.1108/JFMM-03-2015-0021.

Selfridges. (2019). Champagne & Oyster Bar by Caviar House & Prunier. Retrieved from <https://www.selfridges.com/GB/en/features/info/stores/london/store-services/restaurants-bars/champagne--oyster-bar-by-caviar-house--prunier/>.

Solomon, M. R. (2009). *Consumer behavior: Buying, having, and being* (8th ed.). Upper Saddle River, NJ: Pearson Education.

Srinivasan, S. R., & Srivastava, R. K. (2010). Creating the futuristic retail experience through experiential marketing: Is it possible? An exploratory study. *Journal of Retail & Leisure Property*, *9*(3), 193–199. Available from https://doi.org/10.1057/rlp.2010.12.

Stein, A., & Ramaseshan, B. (2016). Towards the identification of customer experience touch point elements. *Journal of Retailing and Consumer Services*, *30*, 8–19. Available from https://doi.org/10.1016/j.jretconser.2015.12.001.

The Sybarite Report. (2019, January). The Sybarite Report 2018: The future of the luxury department store, transcendent retail and the role of design. Retrieved from <https://d217ou440bjdpz.cloudfront.net/documents/The-Sybarite-Report-2018_The-Future-of-the-Luxury-Department-Store.pdf>.

Tynan, C., McKechnie, S., & Chhuon, C. (2010). Co-creating value for luxury brands. *Journal of Business Research*, *63*(11), 1156–1163. Available from https://doi.org/10.1016/j.jbusres.2009.10.012.

Van Camp, D. J., Hooker, N. H., & Souza-Monteiro, D. M. (2010). Adoption of voluntary front of package nutrition schemes in UK food innovations. *British Food Journal*, *112*(6), 580–591. Available from https://doi.org/10.1108/00070701011052673.

Varshneya, G., & Das, G. (2017). Experiential value: Multi-item scale development and validation. *Journal of Retailing and Consumer Services*, *34*, 48–57. Available from https://doi.org/10.1016/j.jretconser.2016.09.010.

Varshneya, G., Das, G., & Khare, A. (2017). Experiential value: A review and future research directions. *Marketing Intelligence & Planning*, *35*(3), 339–357. Available from https://doi.org/10.1108/MIP-05-2016-0075.

Voss, K. E., Spangenberg, E. R., & Grohmann, B. (2003). Measuring the hedonic and utilitarian dimensions of consumer attitude. *Journal of Marketing Research*, *40*(3), 310–320. Available from https://doi.org/10.1509/jmkr.40.3.310.19238.

Yakhlef, A. (2015). Customer experience within retail environments: An embodied, spatial approach. *Marketing Theory*, *15*(4), 545–564. Available from https://doi.org/10.1177/1470593115569016.

Yeoman, I. (2012). *2050 – Tomorrow's tourism* (Vol. 55). Channel View Publications.

Yeoman, I., & McMahon-Beatte, U. (2016). The future of food tourism. *Journal of Tourism Futures*, *2*(1), 95–98. Available from https://doi.org/10.1108/JTF-12-2015-0051.

Yin, R. K. (2003). *Case study research: Design and methods* (3rd ed.). Thousand Oaks, CA: Sage.

Organic food shopping in Australia and Bangladesh: the role of perceived organic food value in the attitude–intention–behavior model

6

Parves Sultan[1], M. Muzakkeerul Huda[2], Tasmiha Tarafder[3], Tina Umnia Hussain[2], Ho Yin Wong[4] and Abu Saleh[5]
[1]School of Business and Law, Central Queensland University, Melbourne, VIC, Australia,
[2]Presidency University, Dhaka, Bangladesh, [3]RMIT University, Melbourne, Australia,
[4]Department of Marketing, Deakin University, Burwood, VIC, Australia,
[5]Cranberra Business School, University of Canberra, Australia

6.1 Introduction

Food shopping is defined as the act of shopping not only associated with buying foods but also related to the food shopping experience, including the extent of attitude, enjoyment, and pleasure in food shopping (Islam, Sayeed, Akhtar, Hossain, & Liza, 2018). This definition implies that food shopping research should include an understanding of the antecedents to food buying behaviors. Being staple products, foods are often considered as a low-involvement purchase, where consumers spend less time, money, and effort to buy their daily food products. However, research finds that consumers often find organic food as costly and requires more effort to buy when compared with conventional food items (Misra & Singh, 2016).

Organic foods are costly because they are produced using environmentally sound practices instead of using conventional pesticides or fertilizers made with synthetic ingredients and not processed using industrial solvents, irradiation, chemical food additives, or genetic engineering (Jeong & Jang, 2019, p. 118). Hence, some studies argue that personal values, including health, wellbeing, environment, and animal welfare often drive consumers to purchase organic foods items (Makatouni, 2002; Padel & Foster, 2005).

While some studies find a positive relationship between values, attitudes, and organic food purchase intentions (Aertsens, Verbeke, Mondelaers, & Huylenbroect, 2009; Chen, 2009), other studies did not find such support in their empirical findings (Chen, 2009; Shepherd, Magnusson, & Sjödén, 2005). Research examining the relationships between consumers' experiences (e.g., attitudes and perceived values),

Case Studies on Food Experiences in Marketing, Retail, and Events. DOI: https://doi.org/10.1016/B978-0-12-817792-1.00006-X

affective reactions (e.g., intentions), and consequent response behaviors have not been studied extensively (Steffen, 2012). Organic food research and assessment of consumption behavior mostly used the theory of planned behavior (TPB) and examined either the attitude–intention relationship or intention–behavior relationship (Hassan, Shiu, & Shaw, 2016b). Most of these studies are in developed countries contexts. However, little research examining the relationships between consumer value, attitude, intention, and behavior (VAIB) with the TPB framework and in a cross-country context has been conducted. Research is also limited between a developed market and developing market contexts (Sultan, Wong, & Sigala, 2018). Therefore a cross-country exploration of the VAIB model examining the differences and similarities about how Australian and Bangladeshi organic food consumers' perceived values and attitudes influence shopping intention and behavior and to what extent gender and income differences influence the prediction of intention and behavior would produce valuable findings.

6.2 The theoretical backdrop, research context, and hypothesis

The TPB is a well-adapted theory in organic food and green consumption literature. TPB's behavioral intention is the causal effects of attitudes, subjective norm and perceived behavioral control, and actual behavior is the final outcome of intention and perceived behavioral control (Ajzen, 1991). TPB does not theorize the relationship between attitude and behavior, and neither does it theorize the relationship between value, intention, and behavior. The meta-analytic study of Hassan et al. (2016b) found that most empirical research exploring "attitude–behavior" or "intention–behavior" gaps did not actually explore behavior but stopped at the intention and that research seeking quantifiable evidence about the "intention–behavior gap" has not been conducted. This motivates the current study to explore further empirical evidence between VAIB in a cross-country context.

Organic food research in a cross-country context is scarce. Our current review finds a handful of cross-country studies, and most of the organic food studies, including comparative research, are from the European countries (Hassan, Shiu, & Parry, 2016a; Thøgersen, 2010). For example, von Meyer-Höfer, Olea-Jaik, Padilla-Bravo, and Spiller (2015b) studied structural differences in TPB variables between a mature organic food market (Germany) and an emerging organic food niche market (Chile), and they only considered respondents with high education and income groups in both countries. Their study also included altruistic and egoistic motives as the driving factors for attitudes and intention, though the egoistic motive was insignificant for the Chilean market. The sociodemographic variables, including education and income, are usually assumed to be important predictors of intention and behavior, but von Meyer-Höfer et al.'s (2015b) study did not consider these variables.

A comparative study between Taiwan and Indonesia on consumers' perceived behavior for organic rice was also conducted. The descriptive findings show that Taiwanese and Indonesian consumers have differences in attitudes because of the different sociodemographic profiles of the consumers (Moslehpour, Van Kien, & Danyfisla, 2014). For example, the frequency analyses show that males were found more willing to purchase organic rice than females, however, females were found more knowledgeable about organic rice than males, and that both Taiwanese and Indonesian consumers with high-income are more willing to purchase organic rice than other income categories (Moslehpour et al., 2014). However, their study did not conduct any moderation test with the sociodemographic variables on the estimated linear regression models.

Each country has some unique sociodemographic characteristics (e.g., income, education, gender, etc.) that influence consumers' shopping behaviors due to differences in perceived attitudes and values. Studies found significant differences in attitude, intention, and behavior between developed and developing countries (Moslehpour et al., 2014; von Meyer-Höfer, Nitzko, & Spiller, 2015a). The extant literature also confirmed that consumers' perceived value dimensions (e.g., financial, functional, individual, and social values) toward organic food differs significantly between two developed countries: the United States and Germany (Seegebarth, Behrens, Klarmann, Hennigs, & Scribner, 2016).

The Australian organic market is estimated to be worth $2.4 billion of which two-thirds of the Australian shoppers buy organic food items because of their perceived health values and that the market has grown by 88% since 2012 (Cain, 2018). Based on considerable organic farmland in Australia, the growth of the Australian market in the Oceania region is remarkable (Willer & Lernoud, 2018). Contrastingly the organic food sales form 1% of the total agricultural produce in Bangladesh (Ahmed, 2018). Large organic food retail chains, like "Prakritik Krishi Biponon Kendra (PKBK)" (or Natural Agriculture Marketing Centre) in Dhaka buys organic food items from 300 to 350 farmers from the adjacent districts. PKBK also educates the consumers about the benefits of consuming organic foods besides marketing the produces (Ahmed, 2018).

Organic food is expensive as compared to conventional food items. In addition, factors like consumers' readiness to purchase, income, attitudes toward organic food, perceived organic food values, availability of information, trustworthy certification, and power-distance in decision making in a family structure may play important roles in shopping decisions of organic foods between a developed and a developing country. Consumers' readiness to purchase, degree of development of organic food markets, the levels of maturity, and growth stages of organic food industries in both Australia and Bangladesh are not similar. Nonetheless, current literature finds that there is a lack of organic food research in both countries (Ahmed & Rahman, 2015; Sultan et al., 2018). Therefore a comparison of organic food consumers' VAIB between a developed market and a developing market would provide important findings for the marketers and policymakers for mutual growth.

In summary, there is insufficient evidence in cross-country context to study how culturally different consumers perceive organic food values and attitudes and to what extent these influence their intention to purchase and actual behavior. Thus the present study extends the TPB model and examines: (1) the relationships between value, attitude, and behavior in the VAIB model and (2) the effects of the sociodemographic variables (e.g. gender and income) on the VAIB model in the Australia–Bangladesh contexts.

Personal values in organic food are defined as perceived health and welfare benefits, and proenvironmental attitudes of consumers (Makatouni, 2002; Padel & Foster, 2005). Ahmed and Rahman's (2015) study with 120 Bangladeshi participants found that 85% of the respondents "agreed" that organic food products are healthier, 74.2% believed that organic foods have high nutritional values, 52.5% perceived that organic foods maintain a high level of safety and control, and 60% believed that organic food is free from harmful chemicals. Similarly an Australian study with 500 respondents finds that three-quarters of the participants agreed that organic food is healthier than conventionally grown food due to a lack of pesticide residues (Lea & Worsley, 2005). Although literature finds that consumers' perceived organic food values are similar across cultures, research examining the role of perceived organic food values in consumers organic food purchase intention and actual behavior is missing, particularly in the context of a developed and developing countries. Therefore

- H1: perceived value toward organic food has a positive and significant effect on shopping intention.
- H2: perceived value toward organic food has a positive and significant effect on shopping behavior.
- H3: shopping intention has a positive and significant mediating effect between perceived value toward organic food and shopping behavior.

Ajzen (1991) defined attitude as persons' favorable or unfavorable evaluations of a subject. Within the TPB framework, the most predictable and stronger relationship in organic food research is the relationship between attitude and intention (Armitage & Conner, 2001; Asif, Xuhui, Nasiri, & Ayyub, 2018; Tarkiainen & Sundqvist, 2009). Ajzen (1991) further argues that stronger attitudes, subjective norms, and perceived control lead to a stronger behavioral intention to perform the specific behavior in question. Attitude reflects an individual's preference to perform or not to perform a behavior (Scalco, Noventa, Sartori, & Ceschi, 2017). However, the relationship between attitude and behavior in the organic food context is underrepresented in the literature. Therefore

- H4: attitude has a positive and significant effect on shopping intention.
- H5: attitude has a positive and significant effect on shopping behavior.
- H6: shopping intention has a positive and significant mediating effect between attitude toward organic food and shopping behavior.

Behavioral intention is defined as the consumer's readiness to perform behaviors and can encompass both positive and negative attitudinal outcomes (Sultan & Wong, 2012). A stronger intention to engage in behavior produces actual behavior (Ajzen, 1991; Vermeir & Verbeke, 2008). The intention to behave and the actual behavior arising from that intention are conceptually distinct. Most studies in organic food consumption literature used intention (Hassan, et al., 2016a; Michaelidou & Hassan, 2014); hence, there is insufficient evidence for the intention–behavior relationship. Therefore

- H7: shopping intention has a positive and significant effect on shopping behavior.
  Although females are found to favor organic foods as compared to males (Hempel & Hamm, 2016; Padel & Foster, 2005), research produced mixed results in gender differences. For example, males in Taiwan and Indonesia were found more willing to purchase organic rice than females, and females were found more knowledgeable about organic rice than males (Moslehpour et al., 2014). Research also found that health-conscious men are found to be increasingly interested in organic foods (Davies, Titterington, & Cochrane, 1995; Lockie, Lyons, Lawrence, & Mummery, 2002). Because organic foods are costlier across the globe, research reported that consumers with high income are more willing to purchase organic produce than other income categories (Moslehpour et al., 2014; von Meyer-Höfer et al., 2015a). Therefore
- H8: gender has a significant moderating effect in the VAIB model.
- H9: income has a significant moderating effect in the VAIB model.

6.3 Research method

The survey instruments for each of the constructs were reflective, and adapted from the current literature, including organic food values (five items, Baker, Thompson, & Engelken, 2004), attitude (eight items, Chen, 2009), behavioral intention (three items, Sultan & Wong, 2012, 2014, 2018), and behavior (three items, Singh & Verma, 2017). The research items are presented in Appendix 6.1. A seven-point Likert-type scale was provided for responses. The questionnaire was reviewed, and its face validity confirmed by an expert panel of academic researchers and respondents through a pilot survey.

With over 18,000 online panel-only members, the random survey for Australia generated 1011 completed and valid responses, and for this purpose, a market research company was employed. Due to inadequate Internet facilities in Bangladesh, the paper-based personal survey technique was applied. The survey questionnaires were randomly distributed to 2000 organic food consumers in four big retail chain stores in Dhaka city. Consumers who were interested had completed the questionnaires and returned those to the surveyor. This generated 227 completed and valid responses from the Bangladeshi organic food consumers. The survey results generated 49.5% male and 50.5% female respondents for Australia. For Bangladesh, there were 67% male and 33% female respondents. Compared with females, males in Bangladesh were found playing a dominant role in responding to the research-related questions because females had been found reluctant in this regard (Kunii, Nakamura, Abdur, & Wakai, 2002). In Bangladesh, being a patriarchal high-power distance society, males play a major role in decision making in a family.

Appendix 6.1 Descriptive statistics of the items.

Items	Overall agree (%)	Overall disagree (%)	Mean (median)	Std. deviation
Organic food is good for the environment				
Australia (Q7_5) Bangladesh (POF_5)	76.4 86.4	3.8 3.0	5.30 (5) 5.89 (6)	1.10 1.21
Organic foods are produced more ethically				
Australia (Q7_6) Bangladesh (POF_6)	69.4 73.0	5.4 6.1	5.12 (5) 5.25 (5)	1.15 1.22
Organic foods are free of chemical residual				
Australia (Q7_10) Bangladesh (POF_10)	70.7 76.3	8.7 8.4	5.10 (5) 5.37 (6)	1.20 1.32
Organic foods are not contaminated by chemical				
Australia (Q7_11) Bangladesh (POF_11)	70.9 74.9	9.1 9.7	5.10 (5) 5.28 (6)	1.23 1.40
Organic foods maintain high food safety standards				
Australia (Q7_12) Bangladesh (POF_12)	66.5 77.5	5.8 8.7	5.05 (5) 5.47 (6)	1.13 1.32
Organic food has health promoting effect				
Australia (Q5_1) Bangladesh (PH_1)	70.0 88.2	7.0 6.2	5.05 (5) 5.69 (6)	1.14 1.27
Organic foods help to live a healthy lifestyle				
Australia (Q5_2) Bangladesh (PH_2)	69.5 90.7	6.8 4.8	5.07 (5) 5.83 (6)	1.17 1.17
Organic foods enhance health				
Australia (Q5_3) Bangladesh (PH_3)	67.1 82.0	7.7 6.6	5.01 (5) 5.56 (6)	1.17 1.24
I would recommend purchasing organic food				
Australia (Q10_7) Bangladesh (BI_7)	58.9 87.2	8.8 4.3	4.89 (5) 5.65 (6)	1.27 1.05

(*Continued*)

Appendix 6.1 (Continued)

Items	Overall agree (%)	Overall disagree (%)	Mean (median)	Std. deviation
If I had to buy foods today, I would buy organic foods				
Australia (Q10_8) Bangladesh (BI_8)	50.2 83.7	14.2 3.90	4.60 (5) 5.61 (6)	1.25 1.09
I continue buying organic foods despite having conventional alternative on sale				
Australia (Q10_1) Bangladesh (AB_1)	69.0 80.2	6.1 6.1	5.01 (5) 5.39 (6)	1.14 1.10
I gladly buy more organic foods if I find it				
Australia (Q10_2) Bangladesh (AB_2)	62.3 81.1	11.6 7.50	4.87 (5) 5.42 (6)	1.26 1.18
I check if food products are organic before I purchase them				
Australia (Q10_3) Bangladesh (AB_3)	48.2 80.7	22.6 6.20	4.40 (4) 5.38 (6)	1.35 1.23

The partial least squares structural equation modeling (PLS-SEM) analysis requires at least five cases per parameter estimate (Bentler & Chou, 1987). Thus 65 valid cases are required for 13 items; nonetheless, the larger sample size is always desirable to minimize sampling error (Randall & Gibson, 1990). To establish valid and reliable scales for each of the constructs and to determine the causal relationships for theory confirmation, this study used the PLS-SEM technique to avoid two serious problems, including inadmissible solutions and factor indeterminacy (Fornell & Bookstein, 1982). PLS-SEM is also capable of modeling latent constructs under nonnormal conditions with small to medium sample sizes (Bagozzi & Yi, 1994; Chin, 2010).

6.4 Results

The demographics of the respondents and the descriptive statistics of the items for both Australian and Bangladeshi samples are presented in Table 6.1 and Appendix 6.1, respectively.

Table 6.2 shows the convergent and discriminant validity tests' results for both Australia and Bangladesh. The Cronbach's alpha ($CA > 0.70$) and composite

Table 6.1 Demographic of the respondents.

Variables	Australia	Bangladesh
Gender		
Male Female	501 (49.5%) 510 (50.5%)	152 (67%) 75 (33%)
Income		
High income Low income Consumers' average spending for organic foods in a month Education (year 10–postgraduate)	17.3% (≥A$105,000) 25.2% (≤A $44,999) 87.9% (≤A$100) 97.4%	10.1% (≥BDT105,000) 28.6% (≤BDT44,999) 85.9% (≤BDT150) 95.5%

Table 6.2 Results of convergent and discriminant validity tests.

Constructs—Countries	CA	CR	AVE	01	02	03	04
Shopping behavior							
Australia	0.849	0.909	0.768	**0.887**			
Bangladesh	0.819	0.892	0.735	**0.875**			
Attitude toward organic food							
Australia	0.912	0.934	0.740	0.634	**0.860**		
Bangladesh	0.833	0.882	0.604	0.429	**0.777**		
Shopping intentions							
Australia	0.871	0.940	0.886	0.828	0.643	**0.941**	
Bangladesh	0.712	0.872	0.774	0.694	0.462	**0.880**	
Perceived value toward organic food							
Australia	0.941	0.962	0.894	0.623	0.596	0.632	**0.946**
Bangladesh	0.856	0.911	0.774	0.344	0.425	0.346	**0.880**

CA, Cronbach's alpha; *CR*, composite reliability; *AVE*, average variance extracted.

reliability (CR > 0.50) values suggest good reliability and internal consistency for the constructs. Convergent validity was ensured by examining the outer loadings (factor loadings > 0.70) of the items of each of the constructs (see Figs. 6.1 and 6.2). All the items' loadings were satisfactory for both Australia and Bangladesh models, and those were found statistically significant (*t*-value > 1.96). However, six items from the three constructs were eliminated due to poor outer loadings (<0.40) (Hair, Hult, Ringle, & Sarstedt, 2014). These deleted items were self-consciousness

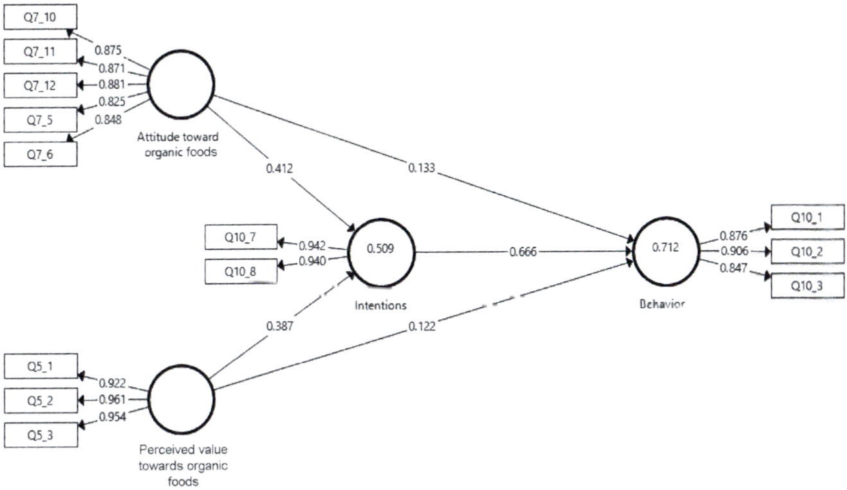

Figure 6.1 The structural model (Australia).

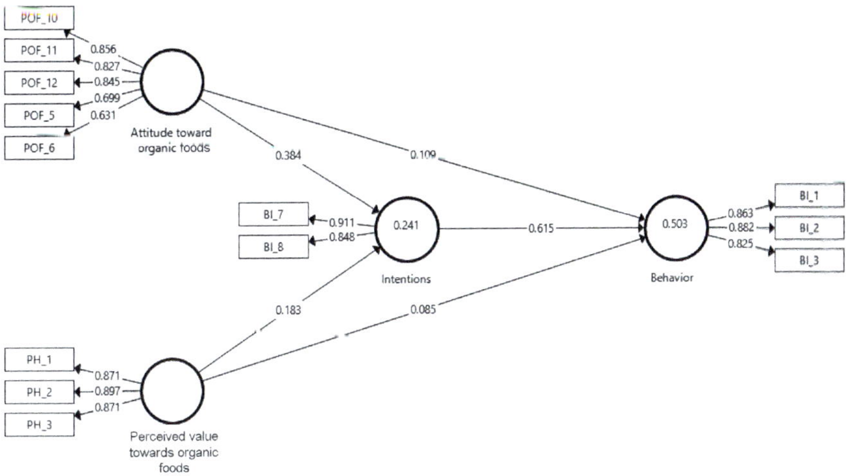

Figure 6.2 The structural model (Bangladesh).

about health (value), maintaining healthy work-life balance (value), superior quality (attitude), organic foods are in fashion (attitude), shopping is joyful (attitude), and preference of organic over conventional food items (intention).

The Fornell–Larcker criterion was used to ensure discriminant validity, where the square root of the average variance extracted (AVE) was compared with the coefficient of the correlation values of a paired construct (Fornell & Larcker, 1981; Hair et al., 2014). Table 6.2 shows that the square root of the AVEs of all constructs (the diagonal values in bold) is greater than the highest correlation value for other constructs. Thus discriminant validity is confirmed for all the constructs. The cross-loadings of items were also examined, where item loadings for their constructs are relatively higher than loadings for the other constructs, and the results confirm the discriminant validity.

The coefficient of determination (R^2) shows the predictive validity of the structural model on a scale between 0 and 1. An R^2 value of ≥ 0.2 indicates a good predictive power (Rodríguez-Entrena, Salazar-Ordóñez, 2013), though the lowest recommended level is ≥ 0.1 (Falk & Miller, 1992). Fig. 6.1 (Australia) shows that behavioral intention explains 50.9% of the variance of attitude and perceived value for organic foods, and together, they explain 71.2% variance in the actual behavior. Similarly Fig. 6.2 (Bangladesh) shows that behavioral intention and actual behavior explains 24.1% and 50.3% of the variance in dependent constructs. The goodness-of-fit (GoF) for the model is the geometric mean of the average communality (outer model) and the average R^2 (inner model). A value greater than 0.36 demonstrates an overall acceptable fit (Hair et al., 2014). Thus the GoF for the Australian and Bangladeshi models are 0.7084 and 0.5178, respectively. Overall, both models demonstrated good explanatory power with an acceptable fitness level.

Table 6.3 shows the hypotheses test results with 2000 bootstrap samples within a 95% confidence interval. In the context of Australia, the study found support for all of the causal hypotheses. Although the results did not find support for two hypotheses in the Bangladesh context, and these are the direct causal effects between attitude, perceived value and actual shopping behavior (Table 6.3), the results in Table 6.4 show that behavioral intention has "full mediation" effects between attitude, perceived value and actual shopping behavior in the context of Bangladesh. However, the partial mediation effects of intention are found significant in the context of Australia.

Table 6.5 shows the multigroup moderation tests with gender (male vs female) and income (high vs low). The multigroup analysis (MGA) results for Australia

Table 6.3 Hypothesis test results.

Causal hypothesis	Australia (β-values)	Bangladesh (β-values)
H1: Perceived value toward organic food has a positive and significant effect on shopping intention	**0.387*****	**0.183*****
H2: Perceived value toward organic food has a positive and significant effect on shopping behavior	**0.122*****	**0.085**
H4: Attitude toward organic food has a positive and significant effect on shopping intention	0.412***	0.384***
H5: Attitude toward organic food has a positive and significant effect on shopping behavior	0.133***	0.109
H7: Shopping intention has a positive and significant effect on shopping behavior	0.666***	0.615***

***$P<.001$. **$P<.01$. *$P<.05$.

Table 6.4 Mediation test results.

Hypothesis	Australia		Bangladesh	
	Direct effect	Indirect effect	Direct effect	Indirect effect
H3: Shopping intention has a positive and significant mediating effect between perceived value toward organic food and shopping behavior	**0.122*** **	**0.257*** **	**0.085**	**0.113*** **
H6: Shopping intention has a positive and significant mediating effect between attitude toward organic food and shopping behavior	0.133***	0.274***	0.109	0.236***

$^{***}P < .001$. $^{**}P < .01$. $^{*}P < .05$.

show that gender and income differences are positive and statistically significant for all paths in the model. However, the results varied for Bangladesh, for example, significant results for male and female are obtained for the attitude→intention, value→intention, and intention→behavior paths. Significant results are also found between high-income and low-income groups for the attitude→intention, intention→behavior paths. The results reveal that high-income and low-income groups have significant moderating effects on value→intention path and value→behavior, respectively. However, the differences of the path coefficients between male and female, and high-income and low-income groups in parametric tests show statistically insignificant results, meaning that the group differences are not substantial though some of the paths have significant effects in the model.

6.5 Discussion

This study aimed to explore the factors affecting the consumers' organic food shopping intentions and actual shopping behavior in a developed and developing countries' context. The theoretical backdrop of this study was the popular TPB theory and particularly the attitude–intention–behavior paths. The novel contributions of the present study have three major folds. First, this study extends the attitude–intention–behavior model by empirically examining the perceived organic food value as the antecedent to the intention and actual behavior. Second, the model was empirically tested and validated for two different countries. Third, mediation of the shopping intention and multigroup analyses of income and gender are performed to better understand the shopping intentions and shopping behaviors of the Australian and Bangladeshi consumers. The discussion of the findings is presented in the following four sections.

Table 6.5 Multigroup moderation test results.

Paths	Gender effect in the model (H8) (β-values)			Income effect in the model (H9) (β-values)		
	Female	Male	Parametric test	High income	Low income	Parametric test
Attitude toward organic food → shopping behavior						
Australia Bangladesh	0.181*** 0.075	0.085*** 0.120	0.096 0.045	0.087*** 0.047	0.162*** 0.131	0.075 0.085
Attitude toward organic food → shopping intention						
Australia Bangladesh	0.429*** 0.473***	0.389*** 0.355***	0.039 0.118	0.429*** 0.341***	0.424*** 0.406***	0.005 0.065
Shopping intention → shopping behavior						
Australia Bangladesh	0.624*** 0.722***	0.704*** 0.578***	0.080 0.144	0.714*** 0.661***	0.630*** 0.500***	0.084 0.161
Perceived value toward organic food → shopping behavior						
Australia Bangladesh	0.119*** 0.052	0.125*** 0.086	0.006 0.034	0.123*** 0.048	0.120*** 0.177***	0.003 0.129
Perceived value toward organic food → shopping intention						
Australia Bangladesh	0.373*** 0.256***	0.399*** 0.143***	0.026 0.113	0.382*** 0.304***	0.373*** 0.118	0.009 0.185

***$P < .001$. **$P < .01$. *$P < .05$.

6.5.1 The roles of attitudes and values in organic food shopping behavior

The current study findings with 227 participants are also consistent with the results of Ahmed and Rahman's (2015) study that finds that 88.2% Bangladeshi organic food consumers perceive that organic foods have health-promoting effect, 82% believed that organic foods enhance one's health, 77.5% perceived that organic foods maintain high food safety standards, and 74.9% believed that chemicals do not contaminate organic foods. Lockie et al.'s (2002) study with 1200 respondents finds that health, natural content, animal welfare and the environment are some of the significant motivating factors to buy organic food products. Their study did not find any significant differences between organic and nonorganic eaters in terms of price, convenience, familiarity, sensory appeal, and religion. Although Lockie et al. (2002) found that income and buying organic foods have a positive correlation, the results did not confirm that income has a significant effect on buying organic foods.

The present study with 1011 Australian participants finds that health is the main reason for buying organic food items. Also the study reveals that consumers generally have positive attitudes toward organic foods because they perceive that organic foods are good for the environment, maintain ethical production and safety standards, and are not contaminated by chemicals. Therefore the present study finds that consumers generally hold positive attitudes toward organic foods. Results from both countries confirm that attitudes and perceived organic food values have significant effects on shopping intention and shopping behavior.

Considering perceived value and attitudes as an antecedent to both intention and behavior within the TPB framework model and examine its effects have earned limited attention in the literature. It is important as previous research demonstrated that consumers intend to buy organic foods because of health-related values (Ahmed & Rahman, 2015; Lockie et al., 2002). Therefore the current study considers the VAIB model and modeled perceived organic food value and attitude as the antecedents to both intention and behavior. Previous research confirmed that TPB could explain 39% and 27% of the variance in behavioral intention and actual behavior, respectively (Armitage & Conner, 2001). A cross-cultural study with an extended TPB also confirmed that the R^2 for purchase intention and behavior of German consumers had been 56% and 53%, respectively, and that for Chile those had been 32% and 17% of variance, respectively (von Meyer-Höfer et al., 2015b). With the addition of perceived organic food value construct in the VAIB model, the present study finds that the R^2 for purchase intention and behavior for Australia are 50.9% and 71.2%, respectively, and that for Bangladesh those are 24.1% and 50.3%, respectively. Thus the present study finds support for adding the perceived value in the VAIB model and contributes to the current literature.

The current study demonstrates that both attitudes and perceived values have positive coefficients and are significant predictors of shopping intentions for both Australia and Bangladesh. The attitude–intention relationship for Australia ($\beta = 0.412$, $P < .001$) is stronger than Bangladesh (($\beta = 0.384$, $P < .001$). Similarly the intention–behavior relationship for Australia ($\beta = 0.666$, $P < .001$) is also stronger than Bangladesh ($\beta = 0.615$, $P < .001$). The study also demonstrates that attitude and perceived value directly and significantly affect the Australian consumers' shopping behaviors. However, the relationships between attitude, perceived value and actual shopping behavior for the Bangladeshi consumers are found statistically insignificant (Table 6.3). The reasons could include the difference in sample sizes for both countries and the difference in the level of consumer awareness and readiness to accept organic foods in both countries.

The study also examined the mediating effects of shopping intention between attitude toward organic food, perceived value toward organic foods and actual shopping behavior. The results with the Australian samples find the "partial-mediational" effects of shopping intention between attitude, perceived value and actual shopping behavior. Contrastingly the results with the Bangladeshi sample did not find any direct relationships between attitude, perceived value and shopping behavior; in fact, those were statistically insignificant (Table 6.4). Interestingly the results show significant "mediational" relationships of shopping intention between organic food, perceived value and actual shopping behavior for the Bangladeshi consumers. Thus the full-mediational effects of shopping intention exist in the model for Bangladesh. This implies that a generally held positive attitudes and perceived values about organic foods in a potential organic food market, such as Bangladesh, may not directly influence shopping/buying behavior as the organic food production, availability, awareness, production, and marketing standards are still in introductory stages in most of the Asian countries (Rehber & Turhan, 2002), including Bangladesh.

6.5.2 The effects of gender and income in organic food shopping behavior

The roles of gender and income differences in the TPB model have inadequate evidence in the literature, and this is particularly absent in a developed–developing countries' context. Thus recent studies have called for more research to find the impact of gender and income differences in the TPB model (Nguyen, Nguyen, & Hoang, 2019; Sreen, Purbey, & Sadarangani, 2018). The current study conducted the MGA with gender (male vs female) and income (high income vs low income), and the results are reported in Table 6.5.

The gender differences in both Australian and Bangladeshi contexts significantly influence all the paths in the model, but the perceived value-shopping behavior path shows insignificant P-value for Bangladesh. Although the results show that the gender differences have significant effects on most of the paths for both Australia and Bangladesh, the effects of the gender differences for each path in the model are found insignificant. This signifies that the gender differences are not substantial for both Australia and Bangladesh. The cross-country findings in this study are consistent with a recent study in the Indian context that examined the gender's effect in TPB model's "attitude–intention" relationship with a sample size of 1100 and did not find any significant differences (Sreen et al., 2018).

The current literature found that income has a positive and a negative causal effect on both green consumption intention and behavior, respectively, in the context of Vietnamese consumers (Nguyen et al., 2019). Our study with the income as a moderating variable showed that attitude–behavior, value–intention, and value–behavior paths were insignificant for Bangladesh; however, all paths were significant for the Australian context.

In the Australian context, the extant literature finds that the most frequently cited purchase barriers include the high relative cost of organic food products, availability, appearance, and lack of adequate information (Sultan et al., 2018). Organic foods are costly as opposed to conventional food items in Australia. Thus personal income plays a major role in organic food shopping intention and actual behavior.

The results also show that income (high income vs low income) has a moderating effect on attitude–intention and intention–behavior paths in the Bangladesh context. This implies that consumers with relatively high income in Bangladesh may have a generally accepted positive attitudes and may be willing to recommend others to purchase organic foods and buy organic foods. However, consumers' perceived values toward organic food and subsequent intention to buy and actual behavior would have no effect due to a change in income. This is because food in Bangladesh is highly contaminated mostly by human-related factors. Almost every day in the daily newspapers, people are alerted of newer methods of adulterating foods through formalin, Carbide, textile colors, DDT, urea, artificial sweeteners, fraud claims, and the like (Ashraf, Joarder, & Ratan, 2018). There is a lack of serious public trustworthiness in food-related claims of the producers and marketers in Bangladesh. Thus this explains why male and female perceptions have no significant differential effect on a few paths in Bangladesh and why Bangladeshi consumers' perceived value toward organic food have no effect on intention to buy and actual behavior. Nevertheless, the effects of the high-income and low-income differences for each path in both models are found insignificant, meaning that the differences are not substantial for both Australia and Bangladesh.

6.6 Conclusion, implications, and future research

This study considered attitude–intention–behavior paths of the TPB model and added perceived organic food value as an antecedent to both shopping intention and shopping behavior. The study samples included organic food consumers from Australia and Bangladesh who have had organic food shopping and consumption experience. Islam et al. (2018) stated that attitudes are one of the major components of food shopping experience, and this is because experience help forming attitudes overtime (Sultan & Wong, 2013). The current study finds that attitude and perceived value are strong and significant predictors of shopping intention for both Australian and Bangladeshi organic food consumers. Similarly intention is also a significant predictor of actual shopping behavior for both countries. Unlike Bangladesh, however, the attitudes and values significantly influence the actual shopping behaviors of the Australian organic food consumers. The findings from two groups of respondents with organic food shopping and consumption experience show that consumers' attitudes toward organic foods refer to environmentally friendly effects, ethical production processes, including chemical and contamination free ingredients and maintaining food safety standards. The findings also reveal that both Australian and Bangladeshi organic food consumers perceive organic food values as having health effects and help maintaining and enhancing healthy lifestyle. Overall, the results of this cross-country findings show that organic food shopping intention and actual shipping behavior are influenced by consumers' shopping attitudes and their health values.

Researchers and practitioners alike notwithstanding of country and cultural differences should focus on building attitudes and perceived values to influence shopping intention and actual shopping behavior. Consumers' attitudes and perceived values are often determined by their shopping/consumption experiences; therefore, this study proposes a 4Cs strategy for enhanced shopping/consumption attitudes and perceived values: Cost, Claims, Certifications and Communications. ***C***ost of production and marketing should remain competitive for enhanced shopping experience as this is the gateway for the new and repeat consumers. Authentic organic Claims regarding the production process, health value, and related information can help enhancing shopping experience and build positive attitudes. Trustworthy and competent Certifications would further enhance shopping experience, build positive attitudes, perceived values; and strengthen shopping intentions and actual shopping behaviors. Appropriate sources and mixes of Communications would educate the market and stakeholders and help sustaining growth of the market and industry. Marketing strategies should include to ensure trustworthy certifications, cold storage, and transportation facilities for organic foods and to minimize consumers' costs by reducing the supply and distribution chains and production and marketing costs through modern technologies. Organic food marketers and producers should also use low-cost promotional and marketing strategies, including market research, direct field to consumers strategies, point-of-purchase promotions, social media promotions, and the like.

The results did not find substantial moderational effects of male versus female and high income versus low income on the VAIB model for both the Australian and Bangladesh contexts. The results of this study show that managers should not differentiate overall communication strategies based on gender and income. This is also supportive of a recent finding for Australia that claimed that gender does not contribute to organic food market segmentation (Sultan et al., 2018).

Australia and Bangladesh can engage in bilateral negotiations and can get mutual benefits from these study findings. Australia being a developed country with advantageous positions in organic farming, certification, and promotion should consider expanding technology transfer and training facilities for the Bangladeshi organic farmers, mainly to take advantages of abundant and inexpensive labors, fertile land and favorable climatic conditions. One of the major limitations of this study is the smaller sample sizes. Future research should consider perceived behavioral control and the subjective norm in account while examining the intention and behavior of the full TPB model and include a larger sample size. Future research should also examine the various roles of modern promotional tools and how those influence VAIB in cross-country contexts.

Author biography

Dr. Parves Sultan is a senior lecturer in Marketing at the Central Queensland University, Australia. His research articles appeared in several scholarly journals, including the *Journal of Brand Management*, *Journal of Business and Industrial Marketing*, *Managing Service Quality (Journal of Service Theory and Practice)*, *International Journal on Disability and Human Development*, *Asia Pacific Journal of Marketing and Logistics*, *Australasian Journal of Regional Studies*, *Quality Assurance in Education*, and *International Journal of Quality and Service Sciences* are worth mentioning. He can be contacted at: p.sultan@cqu.edu.au.

M. Muzakkeerul Huda is an academic staff in the School of Business, Presidency University, Dhaka, Bangladesh. He has published research papers in *International Review of Business Research Papers*, *Indian Journal of Marketing*, *International Journal of Social Business*, *International Journal of Marketing Financial Services and Management Research*, *Journal of Business and Economics*, *Journal of Business and Technology*, and *Journal of Business Administration*. He can be contacted at: muzakkeerrajbd@yahoo.com.

Tasmiha Tarafder is a PhD candidate at the RMIT University, Australia. She had obtained her master by research from the University of Canberra, Australia. She received the Best paper award 2013 and the Highly Commended Award 2012 from the ANZSRAI conferences. She also published in the Proceedings of the 21st Biennial ASAA Conference, Australian National University, Canberra. She published a research paper in the *Australasian Journal of Regional Studies* and a book. She can be contacted at: tasmiha.tarafder@rmit.edu.au.

Tina Umnia Hussain is an academic staff in the School of Business, Presidency University, Dhaka, Bangladesh. She has published a paper in an International Conference in Dhaka and had also coauthored a book chapter on entrepreneurship. She can be contacted at: tina.umnia@gmail.com.

Dr. Ho Yin Wong is a senior lecturer at the Department of Marketing, Deakin University, Burwood Campus, Australia. He has published in the *International Marketing Review*, *Journal of Business and Industrial Marketing*, *Journal of Strategic Marketing*, *Journal of Global Marketing*, *Managing Service Quality*, *Marketing Intelligence and Planning*, *Journal of Brand Management*, *Journal of Product and Brand Management*, and *Asia Pacific Journal of Marketing and Logistics*. He can be contacted at: ho.wong@deakin.edu.au.

Dr. Abu Saleh is an associate professor at the University of Canberra, Australia. He published research papers in *Management Decision*, *Journal of Business Research*, *Australasian Marketing Journal*, *International Journal of Bank Marketing*, and *Asian Social Science*. He can be contacted at: abu.saleh@canberra.edu.au.

References

Aertsens, J., Verbeke, W., Mondelaers, K., & Huylenbroect, G. V. (2009). Personal determinants of organic food consumption: A review. *British Food Journal, 111*(10), 1140–1167.

Ahmed, R., & Rahman, K. (2015). Understanding the consumer behaviour towards organic food: A study of the Bangladesh market. *IOSR Journal of Business and Management, 17*(4), 49–64.

Ahmed, S. *Appetite for organic food on the rise.* (2018). <http://www.theindependentbd.com/post/181034> Accessed 20.06.19.

Ajzen, I. (1991). The theory of planned behaviour. *Organizational Behaviour and Human Decision Process, 50*(2), 179–211.

Armitage, C. J., & Conner, M. (2001). Efficacy of the theory of planned behaviour: A meta-analytic review. *British Journal of Social Psychology, 40*(4), 471–499.

Ashraf, M. A., Joarder, M. H. R., & Ratan, S. R. A. (2018). Consumers' anti-consumption behavior toward organic food purchase: An analysis using SEM. *British Food Journal, 121*, 104–122. Available from https://doi.org/10.1108/BFJ-02-2018-0072.

Asif, M., Xuhui, W., Nasiri, A., & Ayyub, S. (2018). Determinant factors influencing organic food purchase intention and the moderating role of awareness: A comparative analysis. *Food Quality and Preference, 63*, 144–150.

Bagozzi, R. P., & Yi, Y. (1994). Advanced topics in structural equation models. In R. P. Bagozzi (Ed.), *Advanced methods of marketing research* (pp. 1–78). Cambridge: Blackwell.

Baker, S., Thompson, K., & Engelken, J. (2004). Mapping the values driving organic food choice: Germany vs. the UK and UK vs. Germany. *European Journal of Marketing, 38* (8), 995–1012.

Bentler, P., & Chou, C. (1987). Practical issues in structural modelling. *Sociological Methods & Research, 16*(1), 78–117.

Cain, J. *Australian organic market report 2018 excites new investment.* (2018). <http://www.ausfoodnews.com.au/2018/05/14/australian-organic-market-report-2018-excites-new-investment.html> Accessed 20.06.19.

Chen, M. (2009). Attitude toward organic foods among Taiwanese as related to health consciousness, environmental attitudes, and the mediating effects of a healthy lifestyle. *British Food Journal, 111*(2), 165–178.

Chin, W. W. (2010). *How to write up and report PLS analyses – A handbook of partial least squares.* UK: Springer.

Davies, A., Titterington, A. J., & Cochrane, C. (1995). Who buys organic food? A profile of the purchasers of organic food in Northern Ireland. *British Food Journal, 97*(10), 17–23.

Falk, R. F., & Miller, N. B. (1992). *A primer of soft modelling.* Akron, OH: University of Akron.

Fornell, C., & Bookstein, F. L. (1982). Two structural equation models: LISREL and PLS applied to consumer exit-voice theory. *Journal of Marketing Research, 19*(4), 440–452.

Fornell, C., & Larcker, D. F. (1981). Structural equation models with unobservable variables and measurement error: Algebra and statistics. *Journal of Marketing Research, 18*(3), 382–388.

Hair, J. F., Jr, Hult, G. T. M., Ringle, C., & Sarstedt, M. (2014). *A primer on partial least squares structural equation modelling (PLS-SEM).* UK: Sage Publications.

Hassan, L. M., Shiu, E., & Parry, S. (2016a). Addressing the cross-country applicability of the theory of planned behaviour (TPB): A structured review of multi-country TPB studies. *Journal of Consumer Behaviour*, *15*(1), 72–86.

Hassan, L. M., Shiu, E., & Shaw, D. (2016b). Who says there is an intention-behaviour gap? Assessing the empirical evidence of an intention-behaviour gap in ethical consumption. *Journal of Business Ethics*, *136*(2), 219–236.

Hempel, C., & Hamm, U. (2016). How important is local food to organic-minded consumers? *Appetite*, *96*, 309–318.

Islam, M. J., Sayeed, M. A., Akhtar, S., Hossain, M. S., & Liza, A. A. (2018). Consumers profile analysis towards chicken, beef, mutton, fish and egg consumption in Bangladesh. *British Food Journal*, *120*(12), 2818–2831.

Jeong, E., & Jang, S. S. (2019). Price premiums for organic menus at restaurants: What is an acceptable level? *International Journal of Hospitality Management*, *77*, 117–127.

Kunii, O., Nakamura, S., Abdur, R., & Wakai, S. (2002). The impact on health and risk factors of the diarrhoea epidemics in the 1998 Bangladesh floods. *Public Health*, *116*(2), 68–74.

Lea, E., & Worsley, T. (2005). Australians' organic food beliefs, demographics and values. *British Food Journal*, *107*(11), 855–869.

Lockie, S., Lyons, K., Lawrence, G., & Mummery, K. (2002). Eating 'green': Motivations behind organic food consumption in Australia. *Sociologia Rural*, *42*(1), 23–40.

Makatouni, A. (2002). What motivates consumers to buy organic food in the UK? Results from a qualitative study. *British Food Journal*, *104*(3–5), 345–352.

Michaelidou, N., & Hassan, L. (2014). New advances in attitude and behavioural decision-making models. *Journal of Marketing Management*, *30*(5–6), 519–528.

Misra, R., & Singh, D. (2016). An analysis of factors affecting growth of organic food: Perception of consumers in Delhi NCR (India). *British Food Journal*, *118*(9), 2308–2325.

Moslehpour, M., Van Kien, P., & Danyfisla, I. (2014). Differences of customer purchase behavior toward organic rice in Indonesia and Taiwan. *International Journal of Quality and Service Sciences*, *6*(4), 348–368.

Nguyen, H. V., Nguyen, C. H., & Hoang, T. T. B. (2019). Green consumption: Closing the intention-behaviour gap. *Sustainable Development*. Available from https://doi.org/10.1002/sd.1875.

Padel, S., & Foster, C. (2005). Exploring the gap between attitudes and behaviour: Understanding why consumers buy or do not buy organic food. *British Food Journal*, *107*(8), 606–625.

Randall, D. M., & Gibson, A. M. (1990). Methodology in business ethics research: A review and critical assessment. *Journal of Business Ethics*, *9*(6), 457–471.

Rehber, E., & Turhan, S. (2002). Prospects and challenges for developing countries in trade and production of organic food and fibers: The case of Turkey. *British Food Journal*, *104*(3/4/5), 371–390.

Rodríguez-Entrena, M., & Salazar-Ordóñez, M. (2013). Influence of scientific-technical literacy on consumers' behavioural intentions regarding new food. *Appetite*, *60*, 193–202.

Scalco, A., Noventa, S., Sartori, R., & Ceschi, A. (2017). Predicting organic food consumption: A meta-analytic structural equation model based on the theory of planned behavior. *Appetite*, *112*, 235–248.

Seegebarth, B., Behrens, S. H., Klarmann, C., Hennigs, N., & Scribner, L. L. (2016). Customer value perception of organic food: Cultural differences and cross-national segments. *British Food Journal*, *118*(2), 396–411.

Shepherd, R., Magnusson, M., & Sjödén, P. (2005). Determinants of consumer behaviour related to organic foods. *Ambio*, *34*(4–5), 352–359.

Singh, A., & Verma, P. (2017). Factors influencing Indian consumers' actual buying behaviour towards organic food products. *Journal of Cleaner Production*, *167*, 473–483.

Sreen, N., Purbey, S., & Sadarangani, P. (2018). Impact of culture, behavior and gender on green purchase intention. *Journal of Retailing and Consumer Services*, *41*, 177–189.

Steffen, A. (2012). *Critical shopping experiences: Affective reactions and behavioural consequences*. Germany: Südwestdeutscher Verlag für Hochschulschriften.

Sultan, P., & Wong, H. Y. (2012). Service quality in a higher education context: An integrated model. *Asia Pacific Journal of Marketing and Logistics*, *24*(5), 755–784.

Sultan, P., & Wong, H. Y. (2013). Antecedents and consequences of service quality in a higher education context: A qualitative research approach. *Quality Assurance in Education*, *21*(1), 70–95.

Sultan, P., & Wong, H. Y. (2014). An integrated model of service quality, institutional brand and behavioural intentions: The case of a University. *Managing Service Quality*, *24*(5), 487–521.

Sultan, P., & Wong, H. Y. (2018). How service quality affects university brand performance, university brand image and behavioural intention: The mediating effects of satisfaction and trust and moderating roles of gender and study mode. *Journal of Brand Management*, *26*(3), 332–347.

Sultan, P., Wong, H. Y., & Sigala, M. (2018). Segmenting the Australian organic food consumer market. *Asia Pacific Journal of Marketing and Logistics*, *30*(1), 163–181.

Tarkiainen, A., & Sundqvist, S. (2009). Product involvement in organic food consumption: Does ideology meet practice? *Psychology & Marketing*, *26*(9), 844–863.

Thøgersen, J. (2010). Country differences in sustainable consumption: The case of organic food. *Journal of Macromarketing*, *30*(2), 171–185.

Vermeir, I., & Verbeke, W. (2008). Sustainable food consumption among young adults in Belgium: Theory of planned behaviour and the role of confidence and values. *Ecological Economics*, *64*(3), 542–553.

von Meyer-Höfer, M., Nitzko, S., & Spiller, A. (2015a). Is there an expectation gap? Consumers' expectations towards organic: An exploratory survey in mature and emerging European organic food markets. *British Food Journal*, *117*(5), 1527–1546.

von Meyer-Höfer, M., Olea-Jaik, E., Padilla-Bravo, C. A., & Spiller, A. (2015b). Mature and emerging organic markets: Modelling consumer attitude and behaviour with Partial Least Square Approach. *Journal of Food Products Marketing*, *21*(6), 626–653.

Willer, H., Lernoud, J. (2018). The World of organic agriculture: Statistics and emerging trends 2018. In *Organic World.Research Institute of Organic Agriculture (FiBL)* <https://www.organic-world.net/yearbook/yearbook-2018/pdf.html> Accessed 22.10.18.

Promoting organic products in India: the case of Two Brothers Organic Farms

7

Prabhat Kumar[1], *Ardhendu Shekhar Singh*[2] *and*
Dilip Dinkar Ambarkhane[2]
[1]Bharati Vidyapeeth (Deemed University), Pune, India, [2]Symbiosis School of Banking and Finance, Symbiosis International (Deemed University), Pune, India

7.1 Introduction

Awareness on and acceptance of the health benefits of consuming organically cultivated food is growing in the developed and developing world. Customers are now concerned about "where their food comes from" and "how it was produced" (Morgan, Marsden, & Murdoch, 2006). Buoyant by the resultant high demand for such "alternative" food led to the emergence of alternative food networks (AFN) in general and expansion of organic food markets in particular (Maye & Kirwan, 2010). It created opportunity for entrepreneurs who are interested in venturing into supplying authentic organically produced farm products. Realizing the potential of AFN the entrepreneurs of this case, siblings Mr. Satyajit Hange and Mr. Ajinkya Hange, decided to move into organic farming after having a successful stint with corporate world.

There exists a plethora of books and scholarly articles that deal with entrepreneurial outlook of the thriving cult of entrepreneurs who proved themselves in their domain. Nonetheless, in modern era that witnesses fast-paced societal changes, technology disruption, government's dynamic policy making, and ever-changing marketplace, it is imperative to conduct research in the field to keep the findings and propositions contemporary and relevant (Katz & Green, 2007; Kuratko & Hodgetts, 2007; Timmons & Spinelli, 2007). This case study is intended to infuse fresh blood in the entrepreneurship domain in general and organic-farming-entrepreneurship in particular.

7.2 Literature review

The growing population of developed countries led to "industrialization" of the agriculture sector where chemicals, such as petrol-based fertilizers and pesticides,

Case Studies on Food Experiences in Marketing, Retail, and Events. DOI: https://doi.org/10.1016/B978-0-12-817792-1.00007-1

and "monoculture" became catalyst in improving yields at faster pace (Clark & Hobbs, 2016). But later on, scientists like Briton Sir Albert Howard argued that cultivating healthy crops need not depend on soil-fertility annihilating and "environmentally unsustainable" harmonizing macronutrients. As an alternative, they proposed recycling on-farm inputs such as manure, compost, and humus to balance macronutrients of soil and improve the yields and quality of crops (Conford & Dimbleby, 2001). Though the practice of labeling and consuming food products, produced without using chemicals, is in practice for almost a century now, it was first formalized and published in the form of a four-page guideline by the Soil Association in 1967 (Clark & Hobbs, 2016). Certification of organic products helps in determining product specifications, production parameters, and network governance, ultimately leading toward reach to global marketplace (Raynolds, 2004).

Identically in developing countries like India, emphasis was on food security rather than food safety and it led to slow penetration of organic products. Additionally organic food retailers were located in urban areas without good supply chain in place and they charged exorbitant price (Singh, 2006). In recent years, price of agriculture inputs for inorganic farming, that is, fertilizers and pesticides have increased manifold, resulting in increased cost of production, whereas the cost of farming organically is 50% less when compared with the cost of inorganic farming (Manaloor, Srivastava, & Islam, 2016). This has led to the growth of organic food market in the country.

India is witnessing change in the income level of consumers and family structures affecting the attitude of consumers toward organic products. Awareness regarding food safety has increased. Joshi and Hioki (2012) found that there is a link between development of organic markets in India and educational attainment, cost of organic products, number of visits to weekly organic bazaar, and economic recession. Recent studies show that growth in organic product demand in India is not limited to big cities and its penetration is now visible in smaller towns equally. Some of the factors influencing the attitude of consumers toward organic products are perception toward organic food, health consciousness, product information, value for money, accessibility, and trust (Mehra & Ratna, 2014) and perceived healthiness (food phobia) was found to be the most important element in one such study (Nandi, Bokelmann, Gowdru, & Dias, 2016). Consumers have become increasingly concerned about the ill-effects of using chemicals and fertilizers on both human and environmental health. Organic products offer opportunity to consumers for making choices that will reduce the adverse impact on the environment and at the same time reduces health hazards of consumers. With the increasing number of lifestyle diseases such as diabetes, obesity, and heart disorders, consumers have shown growing preference for organic products in India (Rana & Paul, 2017).

Consumers are becoming aware of organic products and are trying to become "citizen-consumer's." A citizen-consumer links ethical practices in the act of eating to social and ecological responsibility. Consumption of organic products is symbolically associated with identity construction and social status manifestation (Grosglik, 2017). The increasing concern of the consumers about the side effects of foods containing chemicals has helped the organic farmers in promoting their products. In India, consumers are showing increasing preference toward the organic products, but there are obstacles such as premium price, reliability of products, and unavailability of retail space for selling these products. The dominance of unorganized players in the food market complicates the issue (Sondhi & Vani, 2007). Furthermore Osswald and Menon (2013) talk about the two major challenges in promoting organic products in India is (1) lack of awareness and (2) the difficulty of meeting consumer needs regarding product range, availability, quality, and price.

As a response to the unethical and standardized practices of the food industry, AFNs is proposed as a new model of organic farming. It includes farmers' market, organic way of food production, direct farm to consumer supply, etc. AFNs try to unite ethical farm food producers and their customers in more personal and shorter way. The offerings of AFNs are regarded as "fresh, diverse, organic, quality, and slow," whereas the network itself is identified as "small-scale, short, traditional, community, local, and embedded" (Edwards, 2016). AFNs are recently blamed for being "incorporated into conventional food networks" to be accommodated by and attuned to big manufacturers and traders (Guthman, 2004). With the change in retail infrastructure, share of organized retailers is increasing in India. In this context, producer organizations will help in linking the large number of small farmers with the modern retail outlets by performing explicit forms of coordination to have their share in value chain (Trebbin, 2014).

Success of an entrepreneurial undertaking depends on myriad of factors. Littunen (2000) listed two characteristics which are commonly found in successful entrepreneurs, namely, intense desire of taking independent decision and embracing risk for perceived benefit. In addition to these two, a few other important themes are proposed by various authors (Buttner & Rosen, 1988; Brandstätter, 1997) that include commitment and determination, leadership ability, drive to find opportunities, tolerance for risk, creativity, self-reliance, and enthusiasm to excel.

Many books are written with the objective to provide advice to first-generation entrepreneurs on personality assessment, evaluating opportunity, performing market research, drafting financial plan, preparing and implementing a business plan, etc. (Churchill & Lewis, 1983; Timmons, Smollen, & Dingee, 1985; Timmons & Spinelli, 2007). Earlier, when Lussier (1995) interviewed 160 successful entrepreneurs to get their advice for would be entrepreneurs, the major advices that emerged were having trade knowledge, readiness to work for extended hours, getting assistance in financial matters, and maintaining sufficient start-up capital.

The existing literatures suggest that agricultural entrepreneurship is a peculiar case in entrepreneurial discipline. In a study that extended to six EU countries, de Wolf and Schoorlemmer (2008) talked about various trends which are giving impetus to the growth of entrepreneurship in agriculture. These trends include market globalization, changing consumer demand, changing supply chain, and changing environment among many other trends. Furthermore in the same study, they proposed six categories of skills required for such entrepreneurs, namely, professional (technical) skills, management (finance, HR, and customer related) skills, opportunity (assessing business opportunities, awareness of threats, innovation, risk management, etc.) skills, strategic (reflection, conceptual, planning and goal setting, etc.) skills, and networking (cooperating, team working, leadership, etc.) skills.

7.3 Methodology

The current study investigates a single case where entrepreneurs, having no prior farming experience, establish a successful organic farming business. Two Brothers Organic Farms (TBOF) was gaining popularity and attracting media attention in the last year. The entrepreneurs, who were acquaintances of the authors, were approached by the authors to get permission for writing the firm's success story in the form of a case. After receiving formal approval, two rounds of interviews were conducted; first a telephone one and later, after a time span of 2 weeks, an in-depth personal interview for 2.5 hours at a mutually agreed place. The purpose of this interview was to understand the motivations for and circumstances of the entrepreneurs' decision to start an organic farming business. After the first interview, information and news articles available in print and electronic media were captured that included, but were not limited to, newspapers, magazines, and periodicals. The information collected through this method was further verified with the entrepreneurs for accuracy.

Two visits were made to the farms to collect first-hand information on the firm's fields, infrastructural facilities, and processing facilities. Information on the firm's dairy unit such as methods and techniques used in rearing mammals in natural environments were collected first hand. Furthermore the mechanisms through which cow dung and cow urine are processed before being used as fertilizer for the soil were recorded. People involved in maintaining these cattle were also interviewed for deeper insight into the business of organic farming.

The information provided by the entrepreneurs were documented in the form of audiorecording and at some instances, especially during the farm visit, it was videorecorded. Smart mobile phones with high pixel cameras were used as recording instrument. Later on, all the recordings were transcribed and the parts of interview, that authors were planning to include in the case, were sent to the entrepreneurs through email for deletion, addition, or corrections. After receiving approval from the entrepreneurs, information was used in relevant parts of the study.

7.4 The case

TBOF is an enterprise founded by two siblings Mr. Satyajit Hange and Mr. Ajinkya Hange. The Hange brothers worked as bankers with reputed large Indian and multinational organizations such as Citibank, HSBC, HDFC, and Kotak before starting their organic farming business. They began the organic farming (while still in the job) on 2 acres or 8 hectare of land in year 2010 and their initial 4 years they operated in loss. In 2015 they to quit their job and do farming as a full-time business. Now, in year 2019, their farming is spread across 35 acres or 14 hectare of land (at Bhodani Village, Pune, Maharashtra, India) and the total turnover of the venture is around INR 30 million (approximately EUR 400,000 at the current exchange rate).

During the initial years (years 2010–15), the firm produced banana, papaya, drumsticks, and sugarcane and sold it at big Indian retail outlets (Star Bazaar, Reliance, Nature's Basket, etc.). Subsequently they moved to direct selling and launched clarified butter (*Ghee*), Jaggery, and Moringa powder in 2016. By 2018 they had a portfolio of 21 products. The firm is currently offering products such as (1) food grains (indigenous varieties of pulse (*Tur*) and Emmer wheat or *Khupli Gehu*); (2) dairy products [clarified butter (*Ghee*) from indigenous variety of cow (*Gir* cows), A2 butter, A2 cheese, A2 buttermilk, and bovine colostrum powder]; and (3) fruits and vegetables (papaya, banana, pomegranate, lemon, etc.). A2 milk is produced by native breeds (without any manipulation) and is healthier and easy to digest. Indian native cow breeds such as Red Sindhi, Gir, Rathi, Shahiwal, and Tharparkar are high-yielding breeds producing A2 milk (Satyanarayana, 2018). The firm attributes the demand from customers as the prime reason behind addition of new products.

These products are marketed under the brand name "AMOREARTH," which is a combination of words "amore" (love) and "earth" (soil) in Spanish. In addition, they have also developed the brand "Freshura," which deals with fresh fruits and vegetables. To increase the credibility, they have obtained certification from "Ecocert" which is a reputed French organic certification body.

The Hange brothers did extensive study and research on organic farming to achieve their vision of growing organic farm products. In their endeavor to learn "organic farming" they came across the benefits of Gir or Gyr cows. Gir is a *Bos Indicus* breed whose origin is in the Gir Forest area of Saurashtra and nearby the area in Gujarat state of India. These cows (Zebu breed in Brazil) can sustain very high temperatures and have high resistance to tropical diseases. They were exported to Brazil more than a hundred years back (Joshi, 1953). During the subsequent period, lack of awareness about their uniqueness has resulted in dwindling of their numbers in India. It is estimated that only 3000–5000 cows of this breed are surviving in India, whereas Brazil has around 4 million (Ferraz & de Felício, 2010).

Initially TBOF purchased three Gir cows and 1 bull and now the firm has more than 60 pure breed *Gir* cows, bulls, and calves. During most part of the day, the cattle are left for "free grazing" and are tied only at the time of milking. To negate the possibility of "inbreeding" they change the bull every 2 or 3 years. The bovine-herd ensures regular supply of pesticide, insecticide, and fertilizer for organic farming. The dung and urine of Gir cows contains microorganism which is helpful in improving the fertility of soil and therefore it is useful as manure (Radha & Rao, 2014). Dung and urine are converted into slurry and is spread across the field. Seeing its importance, the Hange brothers feel that dung and urine are the major products of their cows and milk is only a by-product.

Unlike the prevalent practice of burning the plant residue, at TBOF, dried leaves and twigs are mulched with soil. The mulching process should include both carbon and nitrogen rich substances. Hence, they use sugarcane leaves (high carbon content) and Tur (a kind of pulse) leaves (high nitrogen content) together while making compost through mulching. Mulching helps in keeping soil moist and the temperature low. This improves soil fertility, increases capacity to retain water, and increases the life of microbes. As the farm is located in a drought-prone zone, this intervention increases the efficiency of soil in retaining water which makes a huge difference with regards to water requirement of plants.

During summer, the temperature in this part of the state shoots up to 40°C–45°C and it adversely impacts the quality as well as the life of food crops. To create and facilitate a greenhouse type of microclimate, they have planted around 10,000 bamboo, teak, and Australian pine trees around their farm to create a natural boundary. These trees act as a windshield, help the food plants to withstand the hot summer wave, and assist in lowering the farm temperature. The reduced wind speed facilitates the photosynthesis in plants which otherwise, in high wind speed scenario, will close their cells and stop producing chlorophyll (Červený, Šetlík, Trtílek, & Nedbal, 2009). These additional trees help in improving biodiversity of the farm. To prevent the inflow of chemicals and fertilizers from surrounding fields, they have made trenches around the farm.

Instead of cultivating single crop (monocropping), intercropping or multicropping is practiced at TBOF where more than one variety of crops are grown together, especially a combination of monocot and dicot types. They cultivate four to six varieties of fruits and vegetables on a piece of land along with sugarcane and other crops. Around 80% of these are dicots and rests are monocots. A combination of dicots and monocots helps soil in "fixing carbon" from air (McNear, 2013).

TBOF is situated in a region which receives very little waterfall. Hence, they decided to cultivate only those varieties of crops and fruit/vegetable that require little amount of water. At the same time, the temperature goes very high in summer, because of which they have selected crops, fruits/vegetables which can sustain high amount of sunlight. For example, drumstick, pomegranate, and papaya plants are capable of surviving in high temperature with less amount of water by managing to produce chlorophyll through photosynthesis process (Chandra, Jadhav, & Sharma, 2010; Naik & Thippesh, 2014).

The farm runs on five pillars, namely, multicropping, indigenous seeds, desi cow rearing, mulching, and forest trees. These pillars are based on natural cycle starting with multicropping using indigenous seeds; further inputs such as fertilizers are prepared from desi cow's dung and urine. The cultivation is supported by forest trees which control temperature and ensure biodiversity. All postharvest crop residues are mulched back in the soil (Agriculture Information, 2018).

To transport its products from the farm to the markets they use trucks. They also acquired a refrigerated van to transport their perishable dairy products such as butter, cheese, and buttermilk. Around 80% of the order for their products is generated through their website www.twobrothersfarms.com and their page on amazon.com. They maintain a small warehouse in Pune from where they cater to their online customers. The remaining 20% of the sale is done by directly reaching out to their regular customers in two big cities Pune and Mumbai. Both the brothers accompany the truck/van and travel around 1000 km every week to attend organic farmers' markets. They visit two markets in Pune, one at Pashan on second Sunday of every month and another at Koregaon Park on third Sunday of every month. In addition, they visit two organic farmers' markets in Mumbai (at Dadar every Saturday and Bandra every Sunday). TBOF, which started farming with a "bootstrap capital," has become one of the first farmers in the state directly exporting their organic produce to international locations such as Australia, Dubai, United States, Canada, and Switzerland.

The firm has got some attention from foreign nationals who want to learn their unique approach of organic farming. Interns, travelers, and farmers from more than a dozen foreign countries, including, but not limited to, United States, Canada, Germany, Italy, France, Australia, etc., visited their farm by the year 2019. This practice, in turn, helped TBOF through the crowd sourcing techniques. Buttermilk is a biproduct of processing clarified butter (*Ghee*) and it has limited shelf life. The firm was unable to utilize a large amount of it, as there was very little demand in the local market. An Italian couple, who visited their farm as interns in year 2018, trained them on how to make "Ricotta cheese" out of buttermilk. In the same manner, they learnt to produce high-quality peanut butter from another foreign visitor. In the same year, a young intern from Canada, while working on their farm, helped them in setting an Instagram page for popularizing and sharing the initiatives of TBOF.

Furthermore the Hange brothers decided to increase awareness about organically cultivated food and dairy products. They honor every request to interact with the public; they talk to prospective customers directly and encourage and invite people for "farm visit" to and "farm stay" at their farm. The website of the firm has the provision to accept requests from interested parties for the same. They sell food and strongly believe in the value of their products and propose it as medicine by highlighting the benefits of organic products and ill-effects of chemicals and fertilizers present in normal food products available in the market.

> *Jis se product ki shelf life badhti hai us cheez se aapki self life kam hoti hai (Thing which increases 'shelf life' of products, reduces your 'self life')*
>
> *Satyajit Hange*

They have uploaded many videos on "YouTube" explaining their activities. They have also teamed-up with like-minded people (mostly organic farmers) in Pune and Mumbai through initiatives such as "OrganicWe" in Mumbai and Conscious Community or "CoCo" in Pune. TBOF has acquired more than 250 families (that includes many big celebrities and businessmen) in Mumbai who have become their permanent customers and order firm's product regularly. These influential households and opinion builders, through their very positive social media posts, have helped the firm in becoming a recognizable brand in the organic product space.

Some state governments have recognized the potential of two brothers' organic model and their contribution toward the farming sector. Agricultural departments of the states of Maharashtra, Karnataka, Andhra Pradesh, and Punjab have sent their farmers at TBOF to learn the intricacies of organic farming. The Maharashtra state, where the farm is situated, sent more than 2000 farmers to TBOF and in total 6000 farmers have visited the farm to learn the fundamentals of organic farming. The Hange brothers have also presented their case, delivered lectures, and conducted workshops at many reputed business schools in India like the Indian Institute of Management, Indore. They also reached out to their alma mater "Bishops" which is a reputed convent school in Pune city and have teamed-up with school management to introduce the students to farming. Selected students are now actually involved in organic farming inside their school premise.

Until today, TBOF has not borrowed money from outside and the firm is covering the expenses through its income and investments are made from their own savings. As they grow, the firm is planning to have a state-of-the-art Research and Development laboratory, recruit chemists and scientist, and scale up their operation. They are also planning to establish a processing facility and add more processed food items in their offerings such as papaya leave powder, immunity boosting powder, arthritis powder, and undertake mass production of cheese. For coming up with new products, TBOF is already working with patients and with holistic nutritionists. The Hange brothers also want to handhold farmers and help them to switch to organic farming and connect them to the market.

7.5 Analysis

It is visible from the case that TBOF has got success based on three broad factors. First with regards to traits of entrepreneurs, some of the commonly known traits of successful entrepreneurs are very much visible in the Hange brothers' risk taking behavior and their commitment and enthusiasm about farming. When they took a decision to sacrifice their corporate career and follow the less traveled path of being farmers cultivating organic farming, they clearly showed risk taking traits of their character. Once they committed themselves to the venture, they have shown perseverance and devotion toward their decision to become organic farmer. The way they propagate the benefit of organic food products and compare it with medicine reflects the level of enthusiasm for the purpose. At the same time, they are qualified professionals and treat agriculture as business. Second with regards to consumer related issues, TBOF is dedicated toward crating consumer awareness and the Hange brothers visit various educational and training institutions to reach out to larger audiences. The firm operates in niche product market and provides high-quality products desired by the customers. Third they managed market structure related issues in very efficient manner that includes availing the available space (retail chains and Amazon) for organic product as well as creating their own space in the form of farmer's market.

7.6 Managerial implication

This case is intended to assist entrepreneurs as well as entrepreneurship-trainers in demonstrating, understanding, and appreciating the elements of a successful organic food entrepreneurial venture. Additionally the case puts some light on practical difficulties and challenges entrepreneurs face in the early stage of organic food business operation along with the implementation of innovative practices for enhancing customer experience.

Practitioners can get benefit from the experience of TBOF in creating their own channel of distribution such as farmers' market as well as use of existing online channels like Amazon. To promote their idea, the Hange brothers have used channels such as blogs and YouTube which give idea about the promotional activity of an entrepreneurial firm in early stage when they have limited resources at their disposal.

The entrepreneurs' successful adoption of practices such as farm visit and farm stay may have practical implication for practitioners; it ensured enhanced visibility and acceptance of the brand at minimum cost. These practices move one step further in enriching customer experience: customers become a part of value creation. Furthermore consumer experience was enriched by TBOF through educating the consumers about the benefit of organic farm products through various forums; by way of providing the products near customers' home and in an organic environment (farmer's market) and by hearing the experience of other stakeholders on different promotional platforms such as YouTube, blogs to ensure value cocreation. In addition, TBOF plans to conduct more workshops on organic farming, events at harvest time, etc. that may enhance the experience of multiple stakeholders. These experiences may have good value for an entrepreneur planning to start or already started an organic farming business as well for entrepreneurship-trainers in guiding and motivating budding entrepreneurs.

7.7 Limitation

There are several constraints to the study which include, but are not limited to, the scope of this case limited to a single successful organic farming story. Conducting a multicase study on organic farming would have helped in generalizing the findings. Furthermore the findings of this study do not claim to be relevant and applicable to entrepreneurship in the nonorganic farming sector. Hence, readers should be cautious and make appropriate assumptions while using this case in a nonorganic farming context and, especially, in a nonfarming business. At the same time, this story is of an Indian entrepreneurial venture and may not be fully applicable to a non-Indian environment as some information may be indicative of location constraints.

7.8 Conclusions

The case talks about the story of an agricultural entrepreneurial success where two brothers left their corporate career and ventured into organic farming. It gives details on numerous factors that were important and worked as catalyst while they were inching toward a successful and sustainable future. With their commitment, enthusiasm, and hard work they created reputed and enviable position for themselves in the Indian organic farm food market.

The ability to identify and educate the potential customers was one of the major reasons behind the success of TBOF. In this direction they conducted numerous workshops both on the farm and off the farm, wrote their blogs, and answered the questions consumer asked on the social networking sites. They also exploited the existing popular online channels of distribution such as Amazon to reach the customers easily and effectively. In addition, the accessibility to market increased with the formation of farmers' market enabling consumers to come and experience the product before purchase.

Furthermore they initiated and successfully operationalized the practices of farm visit and farm stay. These two practices enabled the firm to reap the benefit of value cocreation and they were able to add new products in their portfolio as a result of such engagements. The encouragement to visit the farm also helped TBOF in increasing visibility, enriching the experience of customers, and educating them about organic farming.

Funding disclosure

The authors are grateful to their employers for granting required leaves to conduct the study. At the same time authors declare that for this case development, no funding was availed from their employer or any other funding agencies.

Acknowledgment

The authors would like to acknowledge the information provided by Satyajit Hange and are grateful to him and TBOF for the same.

References

Agriculture Information. 2018. Two brothers organic farm – To reconnect with the roots bankers return to agriculture with integrated farm near Pune, Maharashtra. Available from <https://www.agricultureinformation.com/two-brothers-organic-farm-to-reconnect-with-the-roots-bankers-return-to-agriculture-with-integrated-farm-near-pune-maharashtra/> Accessed 12.11.18.

Brandstätter, H. (1997). Becoming an entrepreneur—A question of personality structure? *Journal of Economic Psychology*, *18*(2–3), 157–177.

Buttner, E. H., & Rosen, B. (1988). Bank loan officers' perceptions of the characteristics of men, women, and successful entrepreneurs. *Journal of Business Venturing*, *3*(3), 249–258.

Červený, J., Šetlík, I., Trtílek, M., & Nedbal, L. (2009). Photobioreactor for cultivation and real-time, in-situ measurement of O2 and CO2 exchange rates, growth dynamics, and of chlorophyll fluorescence emission of photoautotrophic microorganisms. *Engineering in Life Sciences*, *9*(3), 247–253.

Chandra, R., Jadhav, V. T., & Sharma, J. (2010). Global scenario of pomegranate (*Punica granatum* L.) culture with special reference to India. *Fruit, Vegetable and Cereal Science and Biotechnology*, *4*(2), 7–18.

Churchill, N. C., & Lewis, V. L. (1983). The five stages of small business growth. *Harvard Business Review*, *61*(3), 30–50.

Clark, L. F., & Hobbs, J. E. (2016). Ready-to-use foods (RUFs). *Encyclopedia of Food and Agricultural Ethics*, 1–8.

Conford, P., & Dimbleby, J. (2001). *Origins of the organic movement*. Floris Books.

De Wolf, P., & Schoorlemmer, H. (2007). *Exploring the significance of entrepreneurship in agriculture*. Resarch Institute of Organic Agriculture FiBL, CH-Frick.; 2.

Edwards, F. (2016). Alternative food networks. *Encyclopedia of food and agricultural ethics. Dordrecht: Springer*.

Ferraz, J. B. S., & de Felício, P. E. (2010). Production systems–An example from Brazil. *Meat Science*, *84*(2), 238–243.

Grosglik, R. (2017). Citizen-consumer revisited: The cultural meanings of organic food consumption in Israel. *Journal of Consumer Culture*, *17*(3), 732–751.

Guthman, J. (2004). The trouble with 'organic lite' in California: A rejoinder to the 'conventionalisation' debate. *Sociologia Ruralis*, *44*(3), 301–316.

Joshi, A., & Hioki, K. (2012). Organic bazaar consumers in India: A case study of the IIRD, India, Aurangabad. *Journal of Food, Agriculture and Environment*, *10*(2), 103–106.

Joshi, N. R. (1953). *Zebu cattle of India and Pakistan: An FAO study*. Rome:: Food and Agriculture Organization of the United Nations.

Katz, J. A., & Green, R. P. (2007). *Entrepreneurial small business*. New York: McGrawHill/Irwin.

Kuratko, D. F., & Hodgetts, R. M. (2007). *Entrepreneurship: Theory, process, practice* (7th ed.). Mason, OH: Thomson/South Western Publishing.

Littunen, H. (2000). Entrepreneurship and the characteristics of the entrepreneurial personality. *International Journal of Entrepreneurial Behavior & Research*, *6*(6), 295–310.

Lussier, R. N. (1995). Startup business advice from business owners to would be entrepreneurs. *SAM Advanced Management Journal*, *60*(1), 10–14.

Manaloor, V., Srivastava, D., & Islam, S. (2016). Growth of organic food industry in India. *Agrofor*, *1*(2), 69–76.

Maye, D., & Kirwan, J. (2010). Alternative food networks. *Sociology of Agriculture and Food*, *20*, 383–389.

McNear, D. H., Jr (2013). The rhizosphere-roots, soil and everything in between. *Nature Education Knowledge*, *4*(3), 1.

Mehra, S., & Ratna, P. A. (2014). Attitude and behaviour of consumers towards organic food: An exploratory study in India. *International Journal of Business Excellence*, *7*(6), 677–699.

Morgan, K., Marsden, T., & Murdoch, J. (2006). *Worlds of food: Place, power, and provenance in the food chain*. Oxford University Press.

Naik, B. H., & Thippesh, D. (2014). Fundamentals of horticulture & production technology of fruit crops. *Pomegranate*, *149*, 152.

Nandi, R., Bokelmann, W., Gowdru, N. V., & Dias, G. (2016). Consumer motives and purchase preferences for organic food products: Empirical evidence from a consumer survey in Bangalore, South India. *Journal of International Food & Agribusiness Marketing*, *28*(1), 74–99.

Osswald, N., Menon, M.K., 2013. Organic food marketing in urban centres of India. International Competence Centre for Organic Agriculture (ICCOA).

Radha, T. K., & Rao, D. L. N. (2014). Plant growth promoting bacteria from cow dung based biodynamic preparations. *Indian Journal of Microbiology*, *54*(4), 413–418.

Rana, J., & Paul, J. (2017). Consumer behavior and purchase intention for organic food: A review and research agenda. *Journal of Retailing and Consumer Services*, *38*, 157–165.

Raynolds, L. T. (2004). The globalization of organic agro-food networks. *World Development*, *32*(5), 725–743.

Satyanarayana, K. V. (2018, April 19). The hype over branded A2 milk. Available from <https://www.thehindubusinessline.com/opinion/columns/the-hype-over-branded-a2-milk/article23605826.ece>.

Singh, S. (2006). *Organic produce supply chains in India: Organisation and governance.* Ahmedabad: CMA Publication No. 222, Centre for Management in Agriculture (CMA), Indian Institute of Management.

Sondhi, N., & Vani, V. (2007). An empirical analysis of the organic retail market in the NCR. *Global Business Review*, *8*(2), 283–302.

Timmons, J. A., Smollen, L. E., & Dingee, A. L. (1985). *New venture creation*. Homewood, IL: Richard D. Irwin.

Timmons, J. A., & Spinelli, S. (2007). *New venture creation: Entrepreneurship for the 21st century*. Burr Ridge: IL: Irwin.

Trebbin, A. (2014). Linking small farmers to modern retail through producer organizations – Experiences with producer companies in India. *Food Policy*, *45*, 35–44.

Engineering the experience in restaurants by means of process design and management

8

Dennis Hanson
The Federal University of Mato Grosso do Sul (UFMS), School of Engineering—FAENG, Pioneiros, Brazil

8.1 Introduction

The purpose of this chapter is to discuss the role of business process management (BPM) in the engineering of a positive experience (Johnston & Kong, 2011; Pine & Gilmore, 1998; Sundbo & Sorensen, 2013) in businesses in general and in restaurants particularly. According to Kauppinen-Räisänen, Gummerus, and Lehtola (2013), a memorable experience is what is expected from a restaurant, and amazing food is only part of the picture.

Thus memorable experiences can and should be designed (Alcántara, Artacho, Martínez, & Zamora, 2014; Grefé, 2011) and/or engineered (Johnston & Kong, 2011). From Pine and Gilmore's (1998) concept of the company as a theatrical stage to business process improvement (Zellner, 2011), a lot has been said on the topic.

It is quite understandable that most restaurants, particularly gourmet restaurants targeted in this study, are small businesses, catering to a generally small, affluent, and selective clientele, under the command of a chef, who may also be the proprietor, or at least one of a small group of partners (Hermelin, Hinchcliffe, & Stenbacka, 2017).

To those unfamiliar with BPM and its concepts, "A process is a set of interrelated or interacting activities which transforms inputs into outputs" (International Organization for Standardization, 2015). Inputs can then be considered as people, materials or ingredients, knowledge, method, machines, and environment. Such elements are put to work toward reaching an output or desired result, which could be goods or services (Slack, Chambers, & Johnston, 2010).

In a restaurant there should be at least processes for procurement (purchases), manufacturing (kitchen), and service. Additionally other activities such as hiring and firing, training, payments due, and among others, can be considered.

Considering this characterization, it is no surprise that BPM or any other sophisticated management tool is simply off the grid for gourmet restaurants, once focus is on the product, rather than in the company or in the customer or the market, and strategic advantage is just not considered at all (Colbert, 2002; Powell, 1992, 2001, 2003; Powell & Arregle, 2007).

Case Studies on Food Experiences in Marketing, Retail, and Events. DOI: https://doi.org/10.1016/B978-0-12-817792-1.00008-3

Large restaurant chains, on the other hand, are usually managed by professionals, using every possible managing tool at hand, in the pursuit of healthy levels of success (Ojiako, Maguire, & Guo, 2009; Shimmura, Takenaka, Ishigaki, & Motomura, 2011).

For the few in the Brazilian restaurant business that have heard of it, BPM sounds complicated (Chong, 2007), expensive, and possibly demanding sophisticated software. On closer scrutiny, this concept does not seem to apply. BPM strives to manage business processes aiming at continuous improvement in all kinds of organizations (Recker, 2014; Rosemann & Vom Brocke, 2015; Vom Brocke et al, 2014).

According to Chong (2007), despite the need for some kind of information technology infrastructure, most of the demands for successful BPM in small organizations is cultural—notably a shift from functional management views to process-oriented frameworks. Bider and Perjons (2015) suggest that the choice of software should be determined by the type of processes at hand and that "The level of details in the model should be sufficient for gaining an understanding of the kind of tools that would be appropriate for this process, but no more" (Bider & Perjons, 2015, p. 3).

Putting it in words a chef or a small restaurant owner would understand, take bread baking, for example: there is a list of ingredients in very exact measures and a series of steps that should be taken to obtain the desired results (Kayser, 2013). Change ingredients, measures, or the steps order and one can get disaster. That specific set of steps has been determined by a long series of trials and errors in which errors are part of the experimenters' learning curve, and once an established practice has been set, a new series of experiments should start to achieve a new set of processes that most possibly will bring about new results.

To take the example a bit further, take the ideal process of the Maître D' approaching an arriving customer.

There, this is BPM in a nutshell. The big question is how to put the principle to use in the bigger picture. Although there is an underlying understanding of process and process management, the main difficulty is that usually there is no explicitly written down set of rules to be followed. In the case of small gourmet restaurants, as targeted in this study, the effort should start with a workflow oriented mind (Van der Aalst, Van, & Hee, 2002).

Fig. 8.1 shows a simple workflow diagram of the processes involved in the reception and seating of clients, as is common in many restaurants. Processes can be defined as simply as the illustration, with squares and arrows showing what should happen in which order.

On the other hand, there are free tools for designing and testing processes. More sophisticated tools should only be considered when things start to work and there is an allotted budget for BPM.

What needs to be done is map, or clearly identify, existing processes, thoroughly check if they work optimally, correct whatever does not work, design the best set of processes possible, and make sure they are followed to the letter. McManus (2018) clearly states the importance of "capturing the error," or going to the root cause of

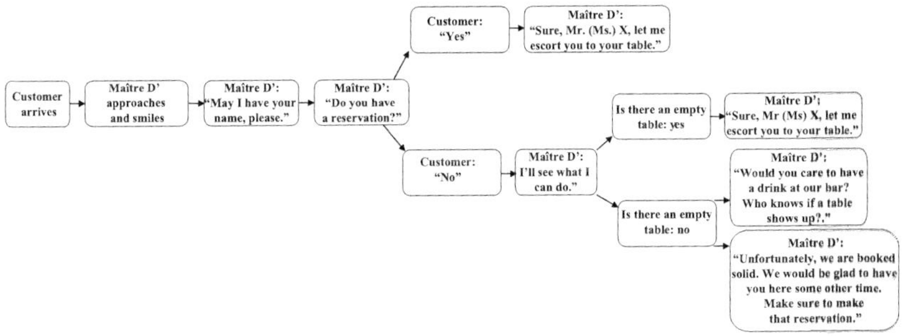

Figure 8.1 Process to approach arriving customers—designed by the author.

errors and defects with the intention of correcting them. In restaurants, as well as in most businesses, poor or nonexistent processes are a major cause of flaws.

8.2 Methodology

A brief introduction presents the reader with recent research on the topics proposed and, due to the intrinsically multidisciplinary nature of the theme, bibliometrics using Boolean operators was chosen to research relevant materials, using keywords such as BPM, business process(es), and restaurants. Nonetheless, of the 286 finds in academic text databases such as the Web of Science and Scopus, only 3 are related to BPM and restaurants. Relevant materials have to be researched by other means, mainly on the Internet.

This research is an "exploratory case study," according to Yin (2014), in a sense that it gathers information, tests methods in the real world to draw questions and point directions, and was applied in a small group of gourmet restaurants in Campo Grande, Mato Grosso do Sul, Brazil. Four examples out of six existing gourmet restaurants were selected to ensure the reader has a clear picture of the gourmet restaurant scenario in town, and whether there is a deliberate focus in using BPM as a means of achieving a memorable experience.

Interviews were conducted with one of the managing partners in two of the restaurants considered and with chefs and proprietors in the other four. There was a nonstructured interview to guide the conversation, but in most part, those interviewed were given free rein to talk as they wished. Interviews lasted from 90 minutes to 2 hours, except in the case of Imakay and 00 Pizza & Pane, which took a full morning, including visits to the "factories."

What the interviews tried to discover was how far each of the restaurants had implemented or used BPM and how did each establishment strive to create a memorable experience for customers.

The criteria considered for best restaurants included food, service, and ambiance quality and are open to debate, as criteria and judgment were subjected to the author's bias. Nevertheless, there is a certain consensus among restaurant goers as to the choice of restaurants, despite there being arguments concerning the inclusion of some typical Brazilian steak houses or *churrascarias*, as they are called locally and one or two other restaurants. *Churrascarias* were deliberately left out as there is nothing really gourmet about them. Their offer is comprised only of a huge amount of barbecued meats, seasoned exclusively in coarse salt, as tradition demands, and a limited buffet of salads and side dishes. Simple fare as French fries is usually greasy and soggy. Other restaurants were left out mainly because the food is not creative enough and service is severely lacking. Two out of the six selected restaurants declined to be interviewed.

8.3 Analysis of the three case studies

In the intention of further enlightening the reader concerning the context and research universe, Campo Grande is generally known as the gateway to the Pantanal, the freshwater wetlands biome that runs through most of the state (Ribas & Schoereder, 2007), with its breathtaking natural scenery and amazing wildlife (far more diversity than the Amazon). The state is also a cattle-growing region, boasting some of the best beef in the country, and freshwater fish are abundant as well. Despite this, gourmet restaurants (Hartmann & Spiller, 2016) are scarce, as well as a demanding clientele, or what Getz and Robinson (2014) would define as foodies, that is, people who view food as an art form, on the same level as painting or drama, as opposed to restaurant goers or even food tourists.

The following case studies showcase the current state of BPM usage in small gourmet restaurants in Campo Grande, Mato Grosso do Sul, Brazil. Coincidentally there are clearly three stages of BPM usage in the four cases provided. All information was provided by the restaurateurs themselves, in semistructured interviews conducted by the author, using open questions and letting the restaurateur speak.

8.3.1 Imakay and 00 Pizza & Pane

Imakay is a nikkey cuisine restaurant, meaning it mixes Peruvian and Japanese cuisine in a very creative manner. Proprietors, chefs, and staff frequently travel to Peru and presently there are two Peruvian chefs working at the restaurant. At present, the restaurant functions at Shopping Campo Grande, the city's most prestigious shopping mall, occupying a 150 m^2 area.

Due to the very small space, there are ongoing plans to move to 600 m^2 in a corner of what is now known as the gourmet district in town.

The space restriction also caused management to be creative. As the administration office occupies a large house, there is an outside area dedicated to food handling, prepreparation, and portioning, with a large pantry and a refrigerated storage room. Every dish on the menu has its own technical register and portions are separated and identified before being sent to the restaurant for cooking and *mis en plât* —or the arrangement of food on the plate to achieve the best visual result possible.

On being interviewed, one of the three partners said they took a couple of years to streamline their processes, and that designing and managing the microprocesses pertaining to kitchen and salon has helped reduce the need for a large staff, stepping down from 35 to 17 persons in the restaurant, as well as reducing investments in training, once each job has its attributions and pertaining processes clearly defined.

"Macroprocesses have never been a problem," says the restaurateur. Purchases, storage, marketing, and the like have been defined at the business' inception. Microprocesses, however, were the fruit of a long learning curve, with hired consultants leading the way and pointing directions. Concerning marketing, the company holds their own marketing company, handling in house and outside clients' demands.

According to the restaurateur, process design and management have greatly reduced personnel turnover, boosted the staff's *ésprit de corps*, and raised customers' satisfaction.

An Imakay branch is due to open in São Paulo, the Brazilian gastronomic capital, and there are plans for another in Belo Horizonte, Brazil's third largest city, in the State of Minas Gerais, famous for its regional food.

The company also owns a few bars and pubs in the city and has recently opened 00 Pizza & Pane, which is discussed next.

00 (Zero, Zero) Pizza & Pane owes its name to one of the types of flour used for the pizzas, imported from Italy, as well as other flours imported from both France and Italy. Their products are sourdough based and everything is made in house in an ample second-floor kitchen, having also a large pantry to hold the several 50 kg flour sacs. The bread and pizza kitchen is fully equipped with all devices necessary, including the large industrial dough mixer and electric fermenting bin. As with Imakay, there is also a large refrigerated room.

There is an interesting fact concerning 00 Pizza & Pane in that sourdough is a novel concept in town, and there had to be a concentrated marketing effort to "educate" the public to accept and enjoy sourdough bread and pizza. Furthermore most of the pizza sold in town is American style, which also means that consumers had to be introduced to Italian style pizza.

Returning to the business aspect, processes at 00 Pizza & Pane have been streamlined to a minimum. There is a small finalization kitchen on the shop floor and customers receive a pager upon entry, choose their topping, and the pizza is rolled down and topped and put into an electric/gas rotating oven imported from Italy. A pizza is ready in 1.5 minutes to be retrieved from the oven by the client himself (Fig. 8.2).

This adds to a thrilling experience, much to the taste of a younger and affluent market, once the pizza prices are a little more expensive than the competition.

The décor is modern rustic, also appealing to the young and affluent, and night hours see the place bursting with clients.

From morning hours to closing time fresh artisanal bread can be picked and paid at a counter in the shop floor. Wines, beers, olive oils, sauces, and condiments can also be bought for takeout, and there are 1 kg bags of French and Italian flours available for sale.

Figure 8.2 Photograph of the imported Italian electric and gas oven used at 00 Pizza & Pane.
Source: Photograph by the author.

The restaurateur says the concept is working fine and there are plans for expansion into other areas of town and to other states as well. From the second-floor bread kitchen, pizza dough balls, and unbaked bread can easily be shipped to the new restaurants.

Following the first two restaurants in the series, which are part of the same company and share a similar philosophy, another interview was conducted with the chef and proprietor of Domus Bistrot.

8.3.2 Domus Bistrot

Located at the heart of the gourmet district in town, Domus Bistrot started out as a French cuisine bistro style restaurant. The restaurant's chef and proprietor, voted best chef in the State of Mato Grosso do Sul in 2018, said he understands the need for properly designed and managed processes. Nevertheless, since the first days of the restaurant's opening in 2016, there has been one crisis after the other—from the major social and economic crisis after president Dilma Roussef's impeachment, to microcrisis in the everyday management of an upscale restaurant.

The chef and proprietor said he has 20 years of experience as a chef and in management, having worked at internationally owned hotels, and owning and managing two other midrange restaurants. Domus is an old dream, bringing the possibility of finally being creative in the catering to an upscale clientele. That is where a steep learning curve began: having come from São Paulo, the country's largest city and gourmet capital, to a smaller capital in a mostly rural state, brought about a series of necessary adaptations to local tastes, even considering the A and B + income bracket target, that is, in the Brazilian market, those whose monthly income is above R$ 10,000 a month (EU 2000) and a 60%/40% female to male rate.

Starting with the menu, in a cattle-breeding state, there was quite a rejection to some of the dishes, particularly those involving shellfish such as scallops, once the ocean is about 2000 km away and local preferences go toward freshwater fish. Beef and stakes are everyone's staple food, and his bestselling dish is risotto and filet mignon on red wine sauce, served quite exclusively well done—the only dish on the menu from day 1. Everything else had to be put to the test. An initial seventy plus dish menu had to be drastically streamlined. Creativity is restricted to a few knowing clients and these dishes are usually off the menu.

A staff of only eight runs the 400 m^2 restaurant from kitchen to salon, which forces everybody to multitask. The chef and proprietor said he is starting to map processes, but this constant multitasking makes it hard to clearly determine who really does what. It is easier in the kitchen, where a staff of three know the ropes and get things done in spite of the proprietor jokingly saying that after 3 months on the job every kitchen boy thinks of himself as an experienced chef. There is a manager who doubles as a *Maître d'* and three waiters who sometimes double as bartenders and cashiers, besides looking after some clients' special demands for dishes off the menu.

All that said, there is an acknowledged need for process design and management, but that has to come after a broader understanding of everyday management, still a difficulty for a chef who also trained in Hotel School.

The final interview was conducted with a fourth restaurant, even considering Imakay and 00 Pizza & Pane are a unit, and brings an interesting sample of what Hermelin et al. (2017) say about gourmet restaurants and restaurateurs.

8.3.3 Les Amis Bistrô

Les Amis is a quaint French-style bistro that seats 80 customers and, as other restaurants in this study, caters for an upscale and discerning clientele. The decor is simple, but sophisticated in its simplicity and the menu is as creative as French cuisine permits. The wine cellar is well stocked and the general atmosphere is cozy and friendly.

When speaking to the husband of the couple that owns and manages the restaurant, the image of the small restaurant described by Hermelin et al. (2017) comes to mind. The product—comprising the setting and decor, the food and the service—is perfect, but management is severely lacking. Macroprocesses such as purchases, inventory management, hiring and training, and the like are carried out rather amateurishly or, to put things straight, are carried out as a very conscientious housekeeper would. Hardly as a business should do it. There is an understanding of a necessary step-by-step for the jobs in the restaurant, but nothing is documented and training is limited to telling employees once what they should do, and if they do not perform, they can be fired. There seems to be rather large employee turnover, as recently a chef and a sommelier were replaced. The new chef who had been recently hired let the bread *levain*, or sourdough, the natural leavening ingredient produced from flour, water, and wild yeasts present in the environment, die for not knowing how to use it. He was a chef and not a baker or *pâtissier*. Now, bread has to be procured in local bakeries, which hardly offer the usual bread fare, quite poor overall. *Levain* baking is limited to three artisanal bakeries, too expensive for most restaurants.

8.4 Findings

As said by the chef and proprietor of Domus Bistrot: "a restaurant is a crazy mixture of a hospital, a factory, and a services company." By this he meant that perfect hygiene and health rules had to be abided, some form of standardization has to be established and service has to be impeccable. Easier said than achieved, and that is where BPM can come to aid the business. When it comes to restaurants, appropriate cleanliness is obviously paramount and there are standards to be enforced in both kitchen and shop (Bai, Wang, Yang, & Gong, 2019; Chow, Alonso, & Douglas, 2010).

A workflow oriented design (Van der Aalst et al., 2002) is many times all that is needed or at least a sound way to go.

As stated above, BPM need not rely on expensive hardware and software. Simple observation of Taylor's and the Gilbreths (Karsten, 1996; Krenn, 2011) times and motion, of a logical, time and cost-effective set of steps is a generally accepted starting point.

The owner of Les Amis has a long learning process in management, the chef from Domus is on the first or second step of the ladder, whereas the owner of Imakay and 00 Pizza & Pane has gone all the way, in spite of admitting processes need to be revised every now and then.

Powell and Arregle (2007) discuss the impact of errors in firm competition and not understanding the importance of properly designed and managed processes both in relation to creating a memorable experience and to profitability is one of such errors.

As stated earlier, there is only one school in town dedicated to training restaurant staff, and most professionals are trained on the job by more experienced staff members. This is more of a problem than a solution—there is quite a lot of entropy, and most professionals are severely lacking in basic skills.

All restaurateurs said that training is usually undertaken in house by more experienced staff and service variability is huge, with a strong impact on customers' experiences. Thus setting up and writing down a simple set of ground rules (Shimmura et al., 2011), much akin to a food recipe, for every member of the staff will suffice both as standard procedure and as guidelines for the training of newly hired staff.

It is understandable that different restaurants may have their particular idiosyncrasics, which may or may not be effective, but the simple understanding of what is going on and how it should go on might be groundbreaking for many of them.

Despite the fact that the research is conducted in Brazil, lessons learned can be adapted to restaurants anywhere in a world in which experiences have become paramount (Johnston & Kong, 2011; Pine & Gilmore, 1998; Sundbo & Sorensen, 2013).

8.5 Managerial implications and conclusions

As stated above, four of the six top restaurants in Campo Grande, State of Mato Grosso do Sul, Brazil, agreed to talk and be featured in this study.

Concerning BPM, there is an evident lesson to be learned: when there is a professional view on management and processes are clearly understood and defined, there is growth and profit. When management is amateurish and there are few or no processes implemented, the best of results are great food, friendly but irregular service, and a nice ambiance, as seen in the study. That would be the case of Les Amis Bistrô and, to a certain extent, Domus Bistrot. Both restaurants keep going with whatever flaws they may have concerning BPM. Both chefs recognize the need for BPM and both agree that it would help in the restaurants' growth.

When one considers Imakay and 00 Pizza & Pane, professional management shows in all aspects of the business, from great food, to the ambience, to the workforce, and the overall experience. Nonetheless, the risk of failure is forever lurking, just waiting for another instance of Brazilian perpetual economic crisis to bring about disaster.

Experience can and should be engineered in the sense that everything related to delighting customers should be carefully thought out. This stands out in the case of Imakay and 00 Pizza & Pane. From procuring the best and freshest ingredients, to décor, to menu design; from kitchen layout and workflow to staff performance; from dish recipes to wine cellar stocking, everything is considered part of the experience and dully noted, even if subconsciously, by customers.

The average life span of bars and restaurants in Campo Grande, Brazil, for that matter, ranges from a few months to a couple of years. The simple fact that these restaurants have survived over 5 years in business cannot be credited to luck. Quality in food, service, and ambiance are responsible for such longevity, but only excellent management, including the understanding and good use of BPM, can bring these restaurants to the next level both in services and in overall experience.

References

Alcántara, E., Artacho, M. A., Martínez, N., & Zamora, T. (2014). Designing experiences strategically. *Journal of Business Research*, *67*(6), 1074–1080.

Bai, L., Wang, M., Yang, Y., & Gong, S. (2019). Food safety in restaurants: The consumer perspective. *International Journal of Hospitality Management*, *77*, 139–146.

Bider, I., & Perjons, E. (2015). Design science in action: Developing a modeling technique for eliciting requirements on business process management (BPM) tools. *Software & Systems Modeling*, *14*(3), 1159–1188.

Chong, Sandy, (2007). Business process management for SMEs: An exploratory study of implementation issues in the western australian wine industry. In: *ACIS 2006 Proceedings — 17th australasian conference on information systems.*

Chow, A., Alonso, A. D., & Douglas, A. C. (2010). Exploring open kitchens' impact on restaurateurs' cleanliness perceptions. *Journal of Retail & Leisure Property*, *9*(2), 93–104.

Colbert, F. (2003). Entrepreneurship and leadership in marketing the arts. *International Journal of Arts and Management*, *6*(1), 30–39.

Grefé, Richard (2011). Experience design is the only design. *Design Management Review*, *22*. Available from https://doi.org/10.1111/j.1948-7169.2011.00153.x.

Getz, D., & Robinson, R. N. S. (2014). Foodies and food events. *Scandinavian Journal of Hospitality and Tourism*, *14*(3), 315–330.

Hartmann, L. H., & Spiller, S. N. A. (2016). The significance of definitional dimensions of luxury food. *British Food Journal*, *118*(8), 1976–1998.

Hermelin, B., Hinchcliffe, G., & Stenbacka, S. (2017). The making of the gourmet restaurateur – Masculine ideology, identity and performance. *NORMA*, *12*(1), 48–64.

International Organization for Standardization. (2015). *Quality management systems – Requirements (ISO standard no. 9001:2015).* Retrieved from <https://www.iso.org/obp/ui/#iso:std:iso:9001:ed-5:v1:en> Accessed 09.10.19.

Johnston, R., & Kong, X. (2011). The customer experience: A road-map for improvement. *Managing Service Quality: An International Journal*, *21*(1), 5–24.

Karsten, L. (1996). Writing and the advent of scientific management: The case of time and motion studies. *Scandinavian Journal of Management*, *12*(1), 41–55.

Kauppinen-Räisänen, H., Gummerus, J., & Lehtola, K. (2013). Remembered eating experiences described by the self, place, food, context and time. *British Food Journal*, *115*(5), 666–685.

Kayser, E. (2013). *Le Larousse du pain – 80 recettes de pains et viennoiseries*. Marseille: Larousse.

Krenn, M. (2011). From scientific management to homemaking: Lillian M. Gilbreth's contributions to the development of management thought. *Management & Organisational History*, *6*(2), 145–161.

McManus, K. (2018). Did you catch it? *ISE Magazine: Industrial & Systems Engineering at Work* (Vol. 50, pp. 16–17). <http://web.a-ebscohost-com.ez51.periodicos.capes.gov.br/ehost/pdfviewer/pdfviewer?vid = 1&sid = 5d10-467f-6471-4c1a-867b-2cc90a2b1873%40sdc-v-sessmgr05>. <https://www.iise.org/ISEMagazine/details.aspx?id = 46699> Accessed 01.23.19.

Ojiako, U., Maguire, S., & Guo, S. (2009). Global operations management during major change: An exploration of industrial practice. *Business Process Management Journal, 15*(5), 816–839.

Pine, B. J., II, & Gilmore, J. H. (1998, July–August). Welcome to the experience economy. Harvard Business Review (pp. 97–105).

Powell, T. C. (1992). Organizational alignment as competitive advantage. *Strategic Management Journal, 13*(2), 119–134.

Powell, T. C. (2001). Competitive advantage: Logical and philosophical considerations. *Strategic Management Journal, 22*(9), 875–888.

Powell, T. C. (2003). Varieties of competitive parity. *Strategic Management Journal, 24*(1), 61–86.

Powell, T. C., & Arregle, J. L. (2007). Firm performance and the axis of errors. *Journal of Management Research (09725814), 7*(2), 59–77.

Recker, J. C. (2014). Suggestions for the next wave of BPM research: Strengthening the theoretical core and exploring the protective belt. *Journal of Information Technology Theory and Application, 15*(2), 5–20.

Ribas, Carla, & Schoereder, José (2007). Ant communities, environmental characteristics and their implications for conservation in the Brazilian Pantanal. *Biodiversity and Conservation, 16*, 1511–1520. Available from https://doi.org/10.1007/s10531-006-9041-x.

Rosemann, M., & Vom Brocke, J. (2015). *The six core elements of business process management*. Berlin, Heidelberg: Springer.

Shimmura, Takeshi, Takenaka, Takeshi, Ishigaki, Tsukasa, Motomura, Yoichi. (2011). Process management in restaurant service — a case study of japanese restaurant chain. In: *International symposium on scheduling*.

Slack, N., Chambers, S., & Johnston, R. (2010). *Operations management* (6th ed). New York: FT Prentice Hall.

Sundbo, J., & Sorensen, F. (2013). *Handbook on the experience economy*. Cheltenham: Edward Elgar.

Van der Aalst, W. M. P., Van., & Hee, K. M. (2002). *Workflow management: Models, methods and systems*. Cambridge: MIT Press.

Vom Brocke, J., Schmiedel, T., Recker, J., Trkman, P., Mertens, W., & Viaene, S. (2014). Ten principles of good business process management. *Business Process Management Journal, 20*(4), 530–548.

Yin, R. K. (2014). *Case study research: Design and methods*. London: SAGE Publications, Inc.

Zellner, G. (2011). A structured evaluation of business process improvement approaches. *Business Process Management Journal, 17*, 203–237.

Ski huts in Austria and their culinary offer: a question of quality versus quantity?

9

Claudia Bauer-Krösbacher, Stephanie Tischler and Dajana Baumgartner
International Business, IMC University of Applied Sciences Krems, Krems, Austria

9.1 Introduction

Winter sports tourism is one of the most important sectors of the tourism industry in general, and even more so in Austria. In the winter season 2018/19, 65 m overnight stays and 17.6 m arrivals were registered in Austria (Austrian National Tourism Office, 2019). Tourism is a key factor of the Alpine country's economy, producing around 50 billion euros every year in the Alpine region and providing 10%–12% of jobs (Disch & Reppe, 2007, p. 5). Although the Alps cover 62% of Austria's territory and 68% of the country's surface area is located above 500 m (Vanat, 2019), Austria's ski industry is highly vulnerable to the effects of climate change. Winter sport tourism resorts have already been adapting their offers to respond to warmer winters (Becker, 2017). Recently many winter sport tourism destinations are trying to diversify their offerings (Hudson & Hudson, 2015, p. 11) and use their culinary and gastronomy offers to position themselves more distinctively (Stöckl, Tischler, Bauer-Krösbacher, Steckenbauer, & Kutsch, 2015, p. 80).

At the same time, changes in consumer behavior in terms of what we eat, how we eat, and where we eat can be observed. Eating out and experiencing the culinary arts of a destination has become a serious leisure activity. However, gastronomy is facing an ongoing change in customer demands, particularly related to current food trends like vegetarian, vegan, or slow food (Rützler, 2015, p. 82; Stöckl et al., 2015, pp. 80–82). Stöckl et al. (2015, pp. 80–82) point out several influencing factors on the food service industry today, such as food intolerances or allergies. Another clear trend in food tourism is the growing importance of healthy, sustainable, and regional products. Consumers increasingly pay attention to the heritage and production of certain products (Everett, 2016, p. 85). Nevertheless, gastronomy seems to be different at 2000 m above sea level (Jurlik, 2017, p. 8). Especially in ski huts, one major challenge is the high number of guests that need to be served at the same time. Ski huts have been criticized by slow food proponents for serving convenience food, which is regarded as low-quality food and unhealthy, especially when combined with sports (Stöckl et al., 2015, p. 80).

Case Studies on Food Experiences in Marketing, Retail, and Events. DOI: https://doi.org/10.1016/B978-0-12-817792-1.00009-5

Many authors have investigated eating out in restaurants as a leisure activity or while being on a holiday (Hall, Mitchell, & Novelli, 2005; Kim, Eves, & Scarles, 2009; Morgan, Hemmington, & Edwards, 2008; Nield, Kozak, & LeGrys, 2000; Pettersson & Fjellströn, 2007). However, little is known about the role of eating for skiers in general and the role of traditional Austrian food. Additionally the investigation of these issues from the gastronomic supply side, more specifically from the perspective of ski hut operators, is limited. Therefore this study aims to:

1. identify the food offer of ski huts in Austria and
2. find out how ski hut operators cope with mass production and the provision of high-quality food.

This contribution begins with a theoretical background on winter sports tourism in Austria, culinary tourism and regional food in Austria as well as culinary experiences at ski huts. The following part is a presentation and discussion of findings from a total of 11 expert interviews with ski hut operators in Salzburg which examined the role of traditional food and the challenges of providing high-quality food at ski huts. Finally conclusions are drawn and recommendations are made for the industry.

9.2 Background of the study

9.2.1 *Winter sports tourism in Austria*

Although the worldwide ski industry faces challenges and the number of skiers in western countries tends to be stagnating or decreasing, globally the industry is very successful (Vanat, 2019). Today, winter sports tourism (skiing and snowboarding) involves around 120 million skiers. From an international perspective, the United States is the leading country in terms of visitors per year with over 57 million tourists. France ranks second with over 56 million visitors, followed by Austria in third place with over 53 million visitors every year (Hudson & Hudson, 2015, p. 14ff).

Currently 67 countries in the world offer equipped outdoor snow-covered ski areas (Vanat, 2019). On a worldwide basis, more than 2000 ski resorts can be identified, with the United States, Japan, and France having the most resorts (Hudson & Hudson, 2015, p. 14). Only Austria and France have more than 10 resorts with a capacity to receive over 1 million skier visits per season. As a result of recent lift-interconnections, Austria possesses the highest number of very large ski resorts. Together with France and the United States, Austria is ranked second in the list of countries with the most lifts, with about 3000 each (Vanat, 2019).

Austria has perfect preconditions for being an attractive winter sport destination due to its topography, climate, and location in central Europe, attractive landscapes, local hospitality, and general enthusiasm toward winter sport among the locals. The Austrian winter sport tourism industry has shown a remarkable development, proven by a steady economic growth for more than half of a century (Arbesser et al., 2008). Since its humble beginnings at the end of the 19th century, winter sports in Austria have been continuously developing. Today, Austria has 254 ski areas with 5 or more lifts (Vanat, 2019). Massive investments have resulted in the most up-to-date lift infrastructure of the industry (Austrian National Tourism Office, 2019).

Already back in the 2000s, many resorts started focusing on restaurants and culinary offers. In the 2000s, a decline in the number of people skiing was observed. Destinations and resorts realized that they had to offer more activities than just skiing, both on- and off-snow. Several alternative offers were created, such as ice-skating, snow-scooting, dog-sledding or paragliding, and restaurants, which became an important component of a ski resort's service offering. Providing culinary experiences during a ski day and after skiing was seen as an integral part of mountain gastronomy (Hudson & Hudson, 2015, pp. 7ff, 97f).

9.2.2 Culinary tourism and regional food in Austria

Culinary tourism—also referred to as food-and-drink tourism—is food-and-drink motivated travel away from home (Everett, 2016, p. 11). This type of tourism can also be regarded as special interest tourism (Hall, Sharples, Mitchell, Macionis, & Cambourne, 2003, p. 11) expressed in visiting a particular food event, a regional producer such as a brewery or cheese maker or simply travelling to a destination which is famous for high-quality food (Hall et al., 2003, p. 7). Several authors point out that food has become an important or major component of today's tourists' itinerary (e.g., Hall et al., 2003, p. 1; Sidali, Spiller, & Schulze, 2011; Yeoman & McMahon-Beatte, 2016, p. 95).

Even though the industry is facing an ongoing internationalization of food and eating habits, the interest in local food and cuisine has been increasing. Local food is defined as food that is produced and consumed within a particular geographical area (Jones, Comfort, & Hillier, 2004, p. 329). Kuznesof, Tregear, and Moxey (1997, pp. 201–202) use a similar definition by emphasizing its strong link to a certain region or place in which it is grown and produced. They further highlight that regional food is food eaten by locals. In addition, they found out that regional food is often linked to old-fashioned food, especially for the younger generation, as they associate the term with traditional ways of preparing home-cooked food, which corresponds to the knowledge and the skills of the older generation (Kuznesof et al., 1997, pp. 201–202). Therefore these authors also highlight the terms "tradition" and "heritage" in relation to regional food. Rützler (2015, p. 75) argues that this trend toward regional products is related to regional and local producers, cultures, and recipes. There is also an increased focus on connecting the regional kitchen with its culture, landscape and ecology (Rützler, 2015, p. 75). Regional food is strongly connected with a particular area since using local products can be seen as its main characteristic. However, it is not just the raw materials used and the area of production that make a regional product regional. The specific ways of food preparation, the recipes used, the people preparing it and the way it is served all contribute to the creation of regional food. Local chefs, for instance, are viewed as experts in the production of local food (Rützler, 2015, p. 204). Therefore employing innovative chefs is essential for the success in the industry (Rützler, 2015, p. 75), as they often use fresh, local, nonindustrialized and organic ingredients as a competitive advantage (Hall et al., 2003, p. 307). Similar to international movements like the "new Nordic cuisine," in Austria a rediscovery of the traditional alpine cuisine can be observed, with a revitalization of specific products and dishes often interpreted in a new way. A key factor in creating "real" regional food is authenticity (Sims, 2009). Kuznesof et al. (1997, p. 203) maintain that personal aspects such as knowledge and experience regarding the product contribute to the authenticity of a regional product. There is an increasing interest in creating regional brands in Austria to strengthen the image of rural destinations such as "Genussregion Österreich" or "AMA Gütesiegel" (Stöckl et al., 2015, p. 82), which certify regional, fresh, and natural products with traceable origin from agricultural production (BMNT, 2019).

9.2.3 Culinary experiences at ski huts

Exploring the offer of a country's regional food and beverages can be regarded as a main motive for travelers, as these expenses make up approximately one-third of the total spending per tourist (Everett, 2016, pp. 13–14). Gastronomy reflects the natural value of a place, its landscape, climate, and culture through unique gastronomic traditions to strengthen the brand image and maintain local identity (Gordin, Trabskaya, & Zelenskaya, 2016, p. 81). Hence, those experiences are essential for providing a positive memory. Each experience created by restaurants affects the potential for building a destinations' brand image (Björk & Kauppinen-Räisänen, 2014, pp. 298–299) and thus people's decision making process leading to the visiting of a particular destination (Sparks, Bowen, & Klag, 2003, pp. 6–7).

With the growth of the skiing industry, the catering facilities in skiing areas have also been changing in terms of numbers, style and size. The range of such facilities comprises traditional small mountain pastures/huts serving traditional Austrian food, big restaurants able to cope with thousands of hungry skiers per day, impressive gourmet style restaurants targeting the more exclusive taste, and fancy bars for the après ski party. Increasingly spectacular architecture is used to provide experiences for customers. Such new hallmarks exist alongside traditional ski huts, which attract tourists due to their perceived charm, ambience and hospitality (Fritz & Fettner, 2017). Only a few add a sort of luxury lifestyle to the initial hospitality, whereas most ski huts still rely on the concept of tradition and quaintness. Many huts offer homemade specialties to provide the feeling of regionality (Fritz & Fettner, 2017). Traditionally there is a strong connection between these huts and agriculture. Most of them are owned by local farmers who use their farm-produced products such as fresh milk, cheese, and meat for gastronomy. Their focus is usually on traditional Austrian food with "Tiroler Gröstl," "Brettljause," "Kaiserschmarrn," and "Apfelstrudel" as some examples. Many of these huts were first only open during the summer but with the success of winter sports started to welcome visitors between Christmas and Easter. Apart from the offered homemade products, they are important places for social gatherings by locals and thereby also provide guests with the opportunity to mingle with the local community.

The following three examples highlight how different winter sport destinations and ski resorts in Austria reacted to these trends:

- The ski resort "Ski Amadé" in Salzburg is trying to position itself as a culinary sport tourism destination and provides a broad culinary offer under the theme "TASTE Ski Amadé." Culinary events, the world's highest farmers' market, wine tastings, Gault&Millau and Falstaff certified restaurants and several initiatives and experiences promoting the culinary specialties of the region are included in this offer (Ski amadé GmbH, 2019).
- The gourmet restaurant "Schaufelspitz," which was awarded 15 points (out of 20) by the Gault-Millau-Guide 2019, is the heart of mountain gastronomy on the Stubai Glacier, Tyrol. It is the world's highest restaurant decorated with two hats. In a noble ambience, sophisticated Austrian cuisine prepared from fresh regional products is served, as well as dishes with international influences (Wintersport Tirol AG & CO Stubaier Bergbahnen KG, 2019).
- Under the motto "Feasting in the world's smallest restaurant," a six-course menu or breakfast is served in one of the gondolas of the "Schönjoch" cable car in Fiss, Tyrol. This offer intends to attract winter sport tourists interested in high-quality local culinary products (Tourismusverband Serfaus-Fiss-Ladis, 2019).

9.3 Methodology

To evaluate the food offer of ski huts in Austria and their operations, a total of 11 expert interviews were carried out with ski hut operators in Salzburg Region. More specifically seven interviews were conducted in Obertauern and four in Zell am See during the winter season 2017. These two destinations represent typical large-scale ski resorts in Austria and rank among the top 20 of the most preferred ski resorts in Austria (Bergfex, 2019).

The specific ski huts were chosen according to criteria such as size, location, and their offer with the ultimate goal of receiving answers from a multitude of different perspectives. All interviews were conducted individually and personally with the respective operators. The interview guide covered 10 main sections: introduction, ski huts in general, logistics, food trends, traditional food, quality, variety, price, unhealthy food, mass production, and closing questions. All interviews were transcribed word by word. Inductive and deductive coding was used to analyze the interviews.

The following parts provide insights into the business of operating a ski hut with a focus on culinary offers. Therefore the results do not cover the perspective of the guests directly, but the operators' views on how to create a culinary offer that meets the needs and wishes of their guests. Original quotes are all direct translations from German into English and are therefore subject to translation bias.

9.4 Findings and discussion

Ski hut operators were first asked what quality means for them and in what way they are able to provide a high level of food quality when catering for a large number of people. For them, food quality depends on numerous factors across the whole food supply chain, but they highlighted that it starts with the selection of the right and reliable supplier so as to work with high-quality raw products. Most of the interviewees consider ski huts to be mass productions when it comes to providing food. Interviewee 1 highlighted the improvement of quality in all sectors:

> *If you are able to manage the gap between mass and quality, the chances to survive in this business are good.*

The challenge is to handle large masses of people who actually have limited time. Ski hut operators argued that their customers are usually stressed, want to have a quick meal and then go back to skiing (Interviewee 3, Interviewee 4, Interviewee 2, and Interviewee 6). Dealing with such large masses seems to be a difficult task for all ski hut operators and their employees. The customers themselves mainly create this stressful situation, as they often want to have lunch at the same time and only take little time for their lunch break. Yet, this seems to depend on the type of guest and the time of year. Interviewee 3 explains the difference between going out to a restaurant as a leisure activity and eating at a ski hut:

> *A ski hut is not a restaurant in which the guest would sit down for a cosy meal and enjoy the evening, but a place to stop just for a bite to eat and drink, maybe also have a coffee and then be back on the slope.*

Furthermore this ski hut operator argues that to be able to cope with this they precook certain dishes in large amounts or quantities, such as "Sugo," to just ladle it out when guests arrive. Nevertheless, all interviewees agreed that mass production in ski huts does not, or only to a slight extent, influence the quality of the food. However, the ultimate goal should be to find a balance between quality and mass production. Interviewee 4 admitted that the food sometimes has to be mass produced, but at the same time, is deeply convinced that quality can still be delivered. This implies that mass production and providing high quality are not mutually exclusive. As a strategy, some ski hut operators for example reduced the choice of dishes in their menus to increase the quality of food as Interviewee 1 explains:

> *(. . .) I rather think that one should reduce the [number of dishes in the] menu. And that is what I did this year (. . .). The choice of dishes is rather small but in return the focus is even more on quality.*

Another key factor for successfully coping with masses of people is suggested by Interviewee 7 when referring to a good working team as a key factor. In addition, the staff also contributes to the quality of the culinary experience for the guest.

A further challenge skiing hut operators face compared with other catering facilities are difficulties with the logistics in the food supply. Because of their location and limited accessibility, specific methods of transportation are needed. Low temperatures and difficult weather conditions can add to this problem. Manzini et al. (2014, p. 2071) stress the importance of logistics in the food supply chain, as this factor also has an impact on the quality of the products. It is argued that the method of transportation and the temperature are major factors which might influence costs and quality (Manzini et al., 2014, p. 2071). Due to the unique location on the mountain, logistics are undoubtedly different. In general, ski huts already start to stock up on dry goods in autumn. This requires a lot of storage space. During the season, depending on their remoteness and accessibility, they receive fresh products usually two to three times a week delivered through skidoos, piste bashers or the gondola. Some interviewees (Interviewee 3, Interviewee 1, Interviewee 2, Interviewee 9, and Interviewee 8) highlighted that the process of delivering food is very time consuming and requires more staff. Interviewee 2 also pointed out that it should not be forgotten that the waste has to be brought down to the valley again, which is another costly step of logistics that is often not considered. Nevertheless, all interviewees stated that the delivery of food has no impact on quality.

Another question of the interviews related to the role of traditional Austrian food. All ski hut operators interviewed by unisono agreed that it is very important to offer traditional Austrian dishes, since this is what tourists expect to get when coming to Austria. Offering traditional Austrian dishes still seems to be the clear focus in Austrian ski huts as highlighted by Interviewee 2:

> *This is our identity. Nothing is better than Austrian cuisine. We are famous all over the world. Absolutely top quality.*

This statement summarizes the main idea of all interviewees. More specifically Interviewee 8 argued that it is almost impossible to operate a ski hut without offering "Kaspressknödel," (beef broth with cheese-bread dumplings) "Kasnockn" (pasta with cheese), or "Tiroler Gröstl" (fried potatoes with fried egg and bacon). In a similar vein, Interviewee 1 maintained the following:

> *Guests do not just associate mountains with skiing or Austria, but also with our culture. We have a proper and long-established culinary culture and that is what they want. This tradition is incredibly important.*

According to Everett (2016, p. 85), a clear trend in food tourism is the growing importance of regionality. Local products are of high importance and a measurement for quality. Operators try to get products/raw material from local suppliers to support the rural community as explained by Interviewee 6:

> *We have many regional suppliers (. . .) for example our butcher (. . .), there we get all our sauces, meat and the famous deer grill sausages (. . .).*

All interviewees claimed that the use of regional products and the support of local businesses was very important to them. Interviewee 4, for instance, argued that regionality stands for quality. Nevertheless, certain limits were mentioned, too. Interviewee 1 underlined the difficulty of defining regionality, especially the problem of defining exactly where regionality ends. He argued that for him Bavaria is regional, too. However, using the concept "Regional 60," where everything within 60 km can be defined as "regional" (Stöckl et al., 2015, p. 87), Bavarian products would not be perceived as regional in Obertauern. He further argued that the important question is who is able to deliver what products in the required amount.

Aspects like the product's origin, sustainability, and environmental impacts appear to be important quality criteria from the consumer's perspective (Botonaki, Polymeros, Tsakiridou, & Mattas, 2006, p. 82). In general, sustainability is becoming more and more important and the attitude of acting sustainably is increasing. All interviewees consider sustainability throughout their whole operation and communicate it to their employees, from ordering the right raw materials to heating, energy and water consumption or waste sorting. Unfortunately there are also limits related to how sustainable a business's products can be, for instance, determined by the way suppliers deliver and package their products.

Customer demand seems to be a major influencing factor for many interviewed ski hut operators. They continuously try to adapt to the changing customer demand to stay successful and satisfy their guests' needs. Interviewee 1 argued that the general consumer behavior has changed radically throughout the past years. Interviewee 4 mentioned that customers demand both traditional and international dishes. It can be said that ski hut operators are trying hard to adapt to the changing customer demand, but that at the same time, it is impossible to offer everything. All ski huts are gradually increasing the offer of vegetarian dishes. However, there is little demand yet for vegan dishes. In addition, due to safety reasons it would be risky to provide vegan cuisine in addition to the usual offer of ski huts. Another interesting fact was stated by Interviewee 5. He pointed out that the demand highly depends on the guests and their origin. People from Vienna, for instance, might demand something completely different than others.

Delivering high-class hospitality in combination with Austrian tradition seems to constitute the major components of the tourist experience. Interviewee 2 also mentioned the location, the view, the cleanliness and the atmosphere as important factors for guests to remember the place. "It is just the happy place," she further added. Interviewee 4 and Interviewee 5 summed this up as the overall package, including infrastructure, hospitality and culinary offer, which needs to be perfect. Interviewee 1 stated that it is necessary to pamper the guests in every possible sense, because at the end of the day they should be satisfied customers. Hospitality, in combination with culinary offer, location and fun, creates a unique memory for all customers visiting Austria and is seen by the Austrian gastronomy as a main factor for success.

Ski huts are often confronted with complaints about excessively high prices. Nevertheless, customers seem to seek high quality, and this does, however, come with a price. The results of this study show that numerous additional costs have to be taken into account regarding the more complex logistics, the higher costs for staff, regular price increases of suppliers, or the short duration of the season—only 5 months. Some customers want to save money on their culinary experiences rather than their leisure time, especially during their skiing vacation.

9.5 Conclusions and practical implications

Culinary experiences, in general, have become an important component of a vacation for tourists. In many places in Austria, the combination of gastronomy and skiing has reached a new level. Results of the interviews suggest that ski hut operators have to deal with numerous challenges compared with a usual restaurant. Findings indicate that not only large numbers of people, but also logistics, staff and changing customer demands are among the main issues ski huts have to tackle nowadays. Although food in ski huts might be generally perceived as mass production, it appears that the quality of the dishes is of a very high standard, supported by the use of local produce.

Eating at a ski hut is a unique experience, which could be further exploited by emphasizing their unique character and the use of local products. For the latter, more information on local suppliers needs to be provided for the customers. To attract more people, off-peak-times breakfast, brunch, or "afternoon snack" could be offered in an innovative, theme-based way. This may help to spread out demand more evenly throughout the day.

Overall, the findings of this study call for a stronger inclusion of the food offers of Austrian ski huts' in marketing. Many Austrian skiing destinations have great potential for exploiting their culinary offers and can only be urged to strengthen their communication efforts in that direction, so as to provide opportunities for differentiation.

With this study, a first step was made to investigate the challenges of ski hut operators with a special focus on quality food versus mass production. Results show that even if the pressure in this industry is very high and characterized by extreme peaks in demand and enormous quantities, it seems that these circumstances do not have a major influence on the quality of the dishes. For practitioners the results can be useful for short- and long-term planning as well as strategy making to adapt to the changing environment. A focus on regional and seasonal products can help not only to create a competitive advantage but also to strengthen the image of ski huts as "ambassadors" of the Austrian cuisine.

This research is limited with regard to its geographical coverage. Furthermore the perspective of customers was not assessed directly. Therefore further research could investigate other skiing destinations in Europe or elsewhere. Using guest surveys or guest interviews to investigate their culinary needs would enhance the understanding of what contributes to their food experiences. Additionally an investigation of different concepts (service vs self-service) should be incorporated in future studies, as this might also affect tourists' experiences.

References

Arbesser, M., Borrmann, J., Felderer, B., Grohall, G., Helmenstein, C., Kleissner, A., et al. (2008). *Ökonomische Bedeutung des Wintersports in Österreich*. Downloaded on April 1, 2017, from <http://www.netzwerk-winter.at/assets/downloads/Netzwerk-Winter_Nachlese-2008/die-oekonomische-bedeutung-des-wintersports-in-oesterreich>.

Austrian National Tourism Office. (2019). Wintersaison 2018/2019 auf Vorjahresniveau: knapp 65 Mio. Nächtigungen (+0,0%) und 17,6 Mio. Ankünfte (+1,3%) im Zeitraum November bis März. Retrieved on May 7, 2019, from <https://www.austriatourism.com/fileadmin/user_upload/Media_Library/Downloads/Tourismusforschung/2019G_Maerz_Hochrechnung_ZusFass.pdf>.

Becker, L. (2017). *Der Weltuntergang im Wintersport fällt aus*. Retrieved on May 7, 2019 from <https://www.ispo.com/maerkte/id_79699794/-der-weltuntergang-im-wintersport-faellt-aus-.html>.

Bergfex GmbH. (2019). *Beliebteste Skigebiete Österreich*. Retrieved on May 17, 2019, from <https://www.bergfex.at/oesterreich/top10/>.

Björk, P., & Kauppinen-Räisänen, H. (2014). Culinary-gastronomic tourism – A search for local food experiences. *Nutrition & Food Science*, *44*(4), 294–309.

BMNT. (2019). *AMA-Gütesiegel*. Retrieved on July 9, 2019, from <https://www.bmnt.gv.at/land/lebensmittel/qs-lebensmittel/AMA-wann-warum.html>.

Botonaki, A., Polymeros, K., Tsakiridou, E., & Mattas, K. (2006). The role of food quality certification on consumers' food choices. *British Food Journal*, *108*(2), 7790.

Disch, D., & Reppe, S. (2007). *Climate change in the alps. Facts – Impacts – Adaptations*. Berlin: Federal Ministry for the Environment, Nature Conservation and Nuclear Safety (BMU) Public Relations Division.

Everett, S. (2016). *Food & drink tourism, principles and practice*. London: Sage Publications.

Fritz, G., & Fettner, F. (2017). *Schihütten*. Retrieved on April 4, 2017, from <http://www.derskiguide.at/wintersport/huetten/abschnitt/2.html>.

Gordin, V., Trabskaya, J., & Zelenskaya, E. (2016). The role of hotel restaurants in gastronomic place branding. *International Journal of Culture Tourism and Hospitality Research*, *10*(1), 81–90.

Hall, C. M., & Mitchell, R. (2005). Gastronomic tourism: Comparing food and wine tourism experiences. In M. Novelli (Ed.), *Niche tourism: Contemporary issues, trends and cases* (pp. 73–88). Oxford: Butterworth-Heinemann.

Hall, C. M., Sharples, L., Mitchell, R., Macionis, N., & Cambourne, B. (2003). *Food tourism around the world. Development, management and markets*. Oxford: Elsevier.

Hudson, S., & Hudson, L. (2015). *Winter sport tourism. Working in winter wonderlands*. Oxford: Goodfellow Publishers Ltd.

Jones, P., Comfort, D., & Hillier, D. (2004). A case study of local food and its routes to market in the UK. *British Food Journal*, *106*(4), 328–335.

Jurlik, M. (2017). Vergessen können Sie vergessen. *Hotel & Touristik*, *3*, 8–11.

Kim, Y. G., Eves, A., & Scarles, C. (2009). Building a model of local food consumption on trips and holidays: A grounded theory approach. *International Journal of Hospitality Management*, *28*(3), 423–431.

Kuznesof, S., Tregear, A., & Moxey, A. (1997). Regional foods: A consumer perspective. *British Food Journal*, *99*(6), 199–206.

Manzini, R., Accorsi, R., Ayyad, Z., Bendini, A., Bendini, M., Bortolini, M., ... Toschi, G. (2014). Sustainability and quality in the food supply chain. A case study of shipment of edible oils. *British Food Journal, 116*(12), 2069–2090.

Morgan, M., Hemmington, N., & Edwards, J. S. A. (2008). From foodservice to food experience? Introduction to the topical focus papers extraordinary experiences in foodservice. *Journal of Foodservice, 19*(3), 151–152.

Nield, K., Kozak, M., & LeGrys, G. (2000). The role of food service in tourist satisfaction. *International Journal of Hospitality Management, 19*(4), 375–384.

Pettersson, A., & Fjellströn, Ch (2007). Restaurants as friends of the family: functions of restaurant visits in everyday life. *Journal of Foodservice, 18*(6), 207–2017.

Rützler, H. (2015). *Food report*. Frankfurt: Zukunftsinstitut.

Sidali, K. L., Spiller, A., & Schulze, B. (2011). *Food, agriculture and tourism – Linking local gastronomy and rural tourism: Interdisciplinary perspectives*. Heidelberg: Springer.

Sims, R. (2009). Food, place and authenticity: Local food and the sustainable tourism experience. *Journal of Sustainable Tourism, 17*, 321–336.

Ski Amadé GmbH. (2019). *Bester Geschmack aus Österreichs Bergen*. Retrieved on May 7, 2019, from <https://www.skiamade.com/de/winter/aktiv-erleben/taste-ski-amade>.

Sparks, B., Bowen, J., & Klag, S. (2003). Restaurants and the tourist market. *International Journal of Contemporary Hospitality Management, 15*(1), 6–13.

Stöckl, A. F., Tischler, S., Bauer-Krösbacher, C., Steckenbauer, G. C. & Kutsch, D. (2015). Kostbares Österreich. A study commissioned by the Bundesministerium für Wissenschaft, Forschung und Wirtschaft. Krems.

Tourismusverband Serfaus-Fiss-Ladis. (2019). *Genussgondel der Schönjochbahn in Fiss*. Retrieved on May 7, 2019, from <https://www.serfaus-fiss-ladis.at/de/Sommerurlaub/Erlebnis-Genuss/Genussgondel>.

Vanat, L. (2019). *International report on snow & mountain tourism*. Retrieved on May 7, 2019, from <https://www.vanat.ch/international-report-on-snow-mountain-tourism>.

Wintersport Tirol AG & CO Stubaier Bergbahnen KG. (2019). *Gourmentrestaurant Schaufelspitz*. Retrieved on May 7, 2019, from <https://www.stubaier-gletscher.com/skigebiet/restaurants/gourmet-restaurant-schaufelspitz/>.

Yeoman, I., & McMahon-Beatte, U. (2016). The future of food tourism. *Journal of Tourism Futures, 2*(1), 95–98.

Food experience design for plant-based restaurants

10

Giovanna Bertella
School of Business and Economics, UiT The Arctic University of Norway, Tromsø, Norway

10.1 Introduction

Although still a minority, the number of people who choose to increase the consumption of plant-based food in Western societies is increasing, and this has important impacts on food businesses, including opportunities to develop new products and experiences (Janda & Trocchia, 2001; Rivera & Shani, 2013; Ruby, 2012; Shani & DiPietro, 2007). Nonetheless, it is only recently that researchers and practitioners have started to investigate and exploit such potential.

Although restaurants that serve exclusively plant-based food might have strict vegetarians as their natural segment, several factors suggest that their services should be attractive also to nonvegetarians. Two recent studies argue that vegetarian hospitality facilities and events need nonvegetarian customers to have a solid income basis, gain legitimacy and spread vegetarianism, an aspect that can be important for the involved entrepreneurs and event organizers (Bertella, 2018a, 2018b).

Moreover, the attractiveness of vegetarian restaurants is important due to the practice of dining out in groups. Nowadays, it is quite common that within a group of people dining out, some of them will be following a plant-based diet. The possibility for vegetarian restaurants to attract nonvegetarians (who usually constitute the majority within groups) relies on the design of particularly memorable experiences. This can be a challenging task due to a tendency to associate vegetarianism to a sort of sacrifice done in the name of ethics and/or health and at the expenses of food pleasure (Cole, 2008).

Due to these aspects—the fast-growing but still limited number of vegetarians and some negativities associated with vegetarianism—it is opportune to explore possible strategies for plant-based restaurants to attract a customer spectrum that includes not just strict vegetarians. This chapter does so, adopting the concepts of neotribe and experience design strategies, and conducting a case study about an Italian café. The aim is to explore possible experience design strategies for plant-based restaurants targeting a customer segment that is not limited to vegetarians.

10.2 Vegetarian neotribes

The social landscape of people who choose to increase the consumption of plant-based food is quite complex. *Vegetarian* is the generic term to indicate people who,

Case Studies on Food Experiences in Marketing, Retail, and Events. DOI: https://doi.org/10.1016/B978-0-12-817792-1.00010-1

based on health and/or ethic-related motivations, choose to exclude or limit animal-derived products from their consumption (Fox & Ward, 2008; Rosenfeld, 2018; Rothgerber, 2014; Ruby, 2012). With regard to food, vegetarians are usually classified as either pure/strict vegetarians, pescovegetarians, lacto-ovo-vegetarians, flexitarians, or reducetarians. Pure/strict vegetarians, also called vegans, exclude all animal-derived food from their diet. The other types of vegetarian follow diets that include some animal-derived food: pescovegetarians eat fish and lacto-ovo-vegetarians include diary and eggs in their diet. Flexitarians and reducetarians, sometimes referred to as occasional vegetarians, consume animal-derived products, but only occasionally.

A term similar to vegetarian is plant-based dieter. A plant-based dieter follows a diet that involves an increased consumption of plant-based foods in comparison to an animal-derived food-centered diet and prefers fresh or minimally processed foods (Lea, Crawford, & Worsley, 2006). Plant-based dieters can coincide with all the aforementioned types of vegetarian, with the only peculiarity being that their choice tends to be health motivated.

The line dividing the various types of vegetarian can be blurry. As suggested by Bertella (2018a), this aspect can be described using the concept of neotribes, that is, fluid social groupings typical of postmodern consumerism (Cova & Cova, 2001; Hardy & Robards, 2015). More precisely Bertella (2018a) identifies three vegetarian neotribes: (1) vegetarian lifestylers, who are particularly motivated and consistent, (2) vegetarians who are in transaction to become vegetarian lifestylers, and (3) vegetarians who follow vegetarianism as a momentary fashion.

The present study adopts the neotribe concept to the plant-based restaurants' target segment. This segment is viewed as fluid, that is, characterized by a spectrum of various motivations and intensity of coherence (Hardy & Robards, 2015). The neotribal fluidity is easily understood in relation to occasional vegetarians, and fluidity can also be related to the other types of vegetarians (Bertella, 2018a). A vegetarian neotribe can comprise vegetarians who, although highly motivated, have inconsistent behavior due to lack of cooking skills possibly deriving from cultural or family related habits (Sobal, Bisogni, Devine, & Jastran, 2006; Van der Kooi, 2010). Another case concerns the transition between the various types of vegetarians, for example, from lacto-ovo-vegetarian to vegan. Moreover, in the present historical time, it can be assumed that most of the vegetarians in Western societies are not born to vegetarian families and have become vegetarians by choice (Leahy, Lyons, & Tol, 2010). Consequently the line dividing nonvegetarians (sometimes referred to as omnivores) and vegetarians is also blurred.

10.3 Experience design strategies

An experience can be understood as "a personal occurrence, often with important emotional significance, founded on the interaction with stimuli which are the products or services consumed" (Carù & Cova, 2003, p. 220). By strategically designing their products and services, suppliers can facilitate the emergence of experiential value relative to various benefits such as entertainment, education, aesthetics, escapism, hedonism, and personal growth (Carù & Cova, 2003; Pine and Gilmore, 1998, 2013).

With regard to the strategic design of experience, Pine and Gilmore (1998) suggest some principles. The first is to thematise the experience: create a sort of story that the customer can live when consuming the offered products and services. Other principles concern the introduction of positive cues that support the chosen theme (e.g., the use memorabilia, the stimulation of all the senses) and the elimination of cues that are in conflict with such theme. Still, according to Pine and Gilmore (1998), the ideal experience occurs in the so-called "sweet spot," a sort of meeting point for various types of experiences, typically educational, aesthetic, escapist, and entertaining.

Recently experiential value cocreation, existing in the process through which the provider and the customer establish relations of trust and, jointly, create reciprocal value, has gained an important role in the design of experiences. According to such position, customers are comparable to partners with whom providers cocreate value, mainly through direct interactions. This idea is applied in various sectors, including food-related sectors such as food industry and hospitality (Boswijk, Thijssen, & Peelen, 2006; Tardivo, Thrassou, Viassone, & Serravalle, 2017).

Within the literature about dining out experiences, some studies identify the following interrelated factors that play a central role in design strategies: food quality, service quality, and atmosphere (Andersson & Mossberg, 2004; Ha & Jang, 2010; Jang & Namkung, 2009; Lee & Crange, 2007; Ryu & Jang, 2008; Soriano, 2002). Food is the core component of the dining out experience design strategy and the following aspects are crucially important: taste, nutritional value, appearance, and coherence with the overall style of the restaurant (Gustafsson, Öström, Johansson, & Mossberg, 2006; Knutson & Patton, 1993).

Atmosphere is specified by some authors using the concept of dinescape, described through six constructs: aesthetics, lighting, ambience, layout, dining equipment, and employees (Ryu & Jang, 2008). The latter element – employees considerably influences the service quality. Here, the findings of Walter, Edvardsson, and Öström (2010) highlight the role of the customer as not "a person in the audience to who something is done" (p. 253) but as an actor who creates the experience through interactions with the provider.

10.4 Plant-based food experiences

Studies concerning plant-based food experiences are scarce: some focus on vegetarian options at generic restaurants and few focus on vegetarian restaurants. Among the first type of studies, a result that emerges is that vegetarians eating at nonvegetarian restaurants tend not to be particularly satisfied. This is found to depend on: the limited and noncreative variety of options, the lack of competence of the waiters, and episodes when traces of animal-derived ingredients are discovered by the customers in their dishes (Shani & DiPietro, 2007; Shani, 2010).

Rivera and Shani (2013) argue that although generic restaurants are aware of the opportunity to attract vegetarians, the developed offers are limited. This is due to the lack of knowledge and competence of restaurants and the general perception of vegetarian food as laborious and dull.

With regard to vegetarian restaurants, Cheng, Lin, and Tsai (2014) report the results of a study about customer preferences in selecting restaurants. The results suggest that food quality is the crucially important element, followed by environmental concern and restaurant design. They also report that an aspect of food quality, namely, its connotation as organic, is the most important factor for customers due to its benefits in terms of health, taste and sustainability. The health aspect is particularly important for mature and female customers. Females constitute the majority of customers who choose vegetarian restaurants. Cheng et al.'s study also suggests that customers might prefer their own native flavors to new ones.

Cagnina, Cicero, and Osti (2017) present an exploratory study about vegetarian restaurants in Italy. Their results confirm that food quality is crucially important, in general and in relation to intentions of revisiting and recommending.

Bertella (2018c) focuses on food experiences at vegan guesthouses. Adopting the entrepreneurs' perspective and a model similar to the dinescape, she identifies the main theme as a story of care and respect toward animals and the environment, and of conflict in relation to dominant norms about animals and food.

10.5 The *Essenza* café case study

Essenza is an Italian café serving plant-based breakfast, lunch, dinner, and snacks. *Essenza* was founded in 2016 by three couples in Grosseto, a town of approximately 82,000 inhabitants located in southern Tuscany (ISTAT, 2018).

10.5.1 Case study identification

The case on which this study relies, the *Essenza* café (Fig. 10.1), was selected based on the expectations about its value in terms of information (Flyvbjerg, 2006). *Essenza* is a valuable case as it is located in a context, Italy, that is particularly relevant to the study of the vegetarianism. In Italy, vegetarians are quite numerous: they constitute approximately the 7% of the adult population, with a minor proportion (0.9%) of vegans (Euripes, 2018). Moreover, Italy is a relevant context for studying vegetarian food experiences due to the importance that food has in the national culture. The data about Italy show also a relatively slow adaption to such trends by restaurants and cafés, and vegetarians are reported to experience various difficulties when dining out (Euripes, 2018).

Essenza, that is known by the researcher who has been an occasional customer since its opening, can be viewed as an extreme case (Flyvbjerg, 2006) due to its success in terms of customer satisfaction. A preliminary search was conducted to gain some information about the customers' evaluation of the café. Customers'

Figure 10.1 *Essenza* plant-based pastries are prepared with high-quality ingredients and displayed elegantly.

reviews on three online platforms, including Google Reviews (212 reviews), Trip Advisor (64 reviews), and the vegetarian food-related platform, Happy Cow (4 reviews), show extremely positive evaluations. This first analysis of the online reviews was based on the evaluation given by the reviewers in the form of stars and labels (excellent, very good, etc.). In this preliminary analysis, it emerged that the average number of stars assigned to *Essenza* on Google is 4.5 out of 5. Of the 64 Trip Advisor, 47 reviews give an excellent (33) or very good (14) valuation. All four reviews on Happy Cow are very positive.

10.5.2 Case study investigation

To learn more about the *Essenza* case, the researcher conducted in-depth interviews with two owners (coded as O1 and O2). Each interview was conducted at the café and lasted approximately 2 hours. The interviews were recorded and transcribed. The interviews followed a guide structured in three sections: factual information about the café (history, vision and strategy, property, staff, partners, and financial aspect), the owners' perspective on the customers (profile: gender, age, and lifestyle), and the factors highlighted in the literature as particularly relevant (food and service quality, atmosphere, and the various components of the dinescape).

The latter factors were used for the content analysis of the customer online reviews collected on the aforementioned platforms. The analysis of the transcripts from the interviews and the online reviews was theory-guided and qualitative: these texts were coded and classified according to the factors emerged as relevant in the dining out experience literature, and the uncoded text was scanned for possible important aspects not included in the theoretical framework of the study (Mayring, 2000). To capture possible new insights, the interviews ended with an open question about other aspects that the owners considered important in relation to the *Essenza* food experience.

10.6 Findings and discussion

10.6.1 The customers: segmenting the market

During the interviews, the owners highlighted that, after some initial challenges, the café has now several loyal customers and continues to attract the curiosity of new ones. The owners stated that, although they have not conducted any survey, they estimate that the vast majority of customers are not vegan. The online reviews show a variety of customer familiarity with plant-based food. The reviews by presumed vegetarians or people with an interest in and some experience with vegetarian restaurants are usually excellent. Two examples of such reviews read as follows:

> *Vegans' paradise. (TA07102018)*
>
> *I'd qualify it as the best vegan restaurant I have ever gone to! (TA21102017)*

Reviews written by presumed nonvegetarians are also very positive, with some commenting on their visit at *Essenza* being their first experience of a vegan restaurant. For example:

> *We are not vegans, but we tried ... and we came back! (TA09072018)*
>
> *First experience at a vegetarian restaurant ... I have to admit that we came here as the other restaurants were closed or full. (...) my husband loves meat so much (...), so I was expecting lots of criticism by him ... but ... everything was excellent. (TA26122018)*

Several reviews recommend *Essenza* to vegetarians and nonvegetarians.

The presence of nonvegetarians among the customers, especially in the cases where such customers did not have any previous experience of vegetarian restaurants, is regarded by the owners as a business success and as an important contribution to a social change that is highly desired, for reasons that include animal and environmental ethical reflections. This is in line with the findings by Bertella (2018a, 2018b), concerning the interest in communicating about and promoting vegetarianism.

The aspect concerning customers being possible nonvegetarians or vegetarians in transition emerged during the interviews. An owner said:

> *We are a vegan café . . . but this does not mean that we are only for vegans! Let's think for a moment: are Chinese restaurants only for Chinese? Well . . . not really . . . quite the contrary! (O2)*

The other owner noted:

> *I used to eat meat and all the rest until not many years ago. I know very well the process . . . looking for options, new ways of eating . . . it is a continuous change. (. . .). We are here to show that the alternative [eating vegan] can be chosen. . . (O1)*

The latter and similar comments suggest that the owners are very aware of the flexibility that characterizes the vegetarian segment. The understanding of this segment by the owners seems to be in line with the neotribe concept proposed by Bertella (2018a).

The owners reported that the vast majority of customers are women. This element, which also emerged in other studies about vegetarianism, can be related to the high percentage of women among vegetarians in Italy (Euripes, 2018). Another explanation might concern the theme of the *Essenza* experience, as presented in the next section and explicitly related to the gender issue, according to the owners.

10.6.2 The Essenza experience: choosing the theme

The theme for the *Essenza* experience can be described as a relaxed, mindful and aesthetically beautiful experience of food pleasure. The elements of relaxation and mindfulness refer to what an owner defined as "the most important of our daily acts, that is eating" (O1). The owners want to offer *Essenza* customers the opportunity to dedicate time to enjoying food, to learning about it and to make choices according to their ideas about health and ethics.

The beauty and pleasure aspects of the *Essenza* experience address what the owners feel is missing in many plant-based restaurants. An owner said that, based on his experience in Italy and in other countries, it is only recently that this type of restaurant has started to properly design the aesthetics around the food experience. The *Essenza* owners have chosen a modern and elegant style. This choice can be related to the local context. As the owners observed, a modern urban café was missing in Grosseto. Grosseto is a small-sized town characterized by an economy based largely on primary sector activities and tourism. Previously residents could experience this type of café only when traveling. Having the possibility to visit a similar place in their hometown seems to be appreciated. This is confirmed by several customers who, according to the owners and as shown in some reviews, comment on the atmosphere of *Essenza*, comparing it to urban centers as New York and London.

With regard to pleasure, the owners observed that some vegan foods available on the market are lacking in taste. This element, added to the lack of aesthetic surroundings that was also mentioned by the owners while commenting on plant-based restaurants they had visited as customers, can make the vegan food dining out experience "a sad experience, comparable to being served tasteless food at the hospital" (O2). The theme of *Essenza* is developed diametrically opposite to such a view.

It can be noted that, in addition to the hedonic aspect, the *Essenza* experience includes several of the potential characteristics of experiences identified in the literature; in particular, aesthetic, educational, personal growth, and escapism. In this sense, the *Essenza* experience can be viewed as gravitating around an ideal "sweet spot." This was commented on by the owners, who said that they want the experience to be "at 360 degrees" (O1). On the webpage, this idea is expressed with the statement: "At *Essenza*, at any time, there is always something to do and to eat."

As mentioned above, the majority of the customers are women and, according to the owners, this can be related to the theme of the experience. According to them, the practice of dedicating time to eating well and relaxing, and the appreciation of elegance might belong more to a feminine approach to life than a masculine one.

Finally it can be noted that the *Essenza* theme is quite different from the theme described by Bertella (2018b), as characterized by a central role of the ethical dimension of eating and a conflictual relationship with the norms of the local context. Several comments from the interviews show that the *Essenza* owners agree on such a way to view veganism. At the same time, they have chosen another direction for the development of the theme for their business and the relative design strategies, as is presented in the next section.

10.6.3 Communicating the theme and facilitating experiential value cocreation

10.6.3.1 Eliminating potential negative cues

The owners have eliminated all the symbolic and physical factors of the dinescape that have references to veganism. The terms vegetarian and vegan are not used to qualify the café, presented with elements that might suggest only covertly the centrality of plant-based food. The words "*Essenza* Love Natural Food" on some of the promotional material highlights the element of care for nonprocessed food, and the logo is a heart with a leaf colored with different shades of green.

Although the food meets the required criteria to be certified as vegan, the owners decided not to apply for this label, suspecting that it might be misunderstood (as "food only for vegans"), and lead to some negative associations due to prejudices and misunderstandings about veganism. The owners observed that vegan food is sometimes confused with vegetarian food but is also associated with being organic, healthy, exotic and "strange." Even more problematic are the associations concerning vegan food as tasteless and lacking variation, and those concerning vegans as extremists. These stereotypes are in line with some comments about vegetarianism and veganism in the literature (Cole, 2008; Rivera & Shani, 2013).

10.6.3.2 Positive cues, cocreation, and memorabilia

Most of the cues used to communicate the theme builds around the critically important factors commented on in the literature; in particular, food quality in relation to taste, nutritional value, and sustainability (organic food) (Andersson & Mossberg, 2004; Cagnina et al., 2017; Cheng et al., 2014; Ha & Jang, 2010; Jang & Namkung, 2009; Lee & Crange, 2007; Ryu & Jang, 2008; Soriano, 2002). Some attention is also observed in the way food is presented (Gustafsson et al., 2006; Ryu & Jang, 2008): simply and elegantly.

With regard to taste, both new tastes and traditional tastes are included. This is done by taking inspiration from ethnic cuisine, especially Indian cuisine, and the traditionally Italian cuisine, and by veganizing traditionally animal-derived foods. Such a combination, particularly evident at lunch, when a buffet with numerous choices is arranged, seems to be appreciated by the customers. Some reviews comment positively about the "innovative" and "creative" food offerings, while others appreciate the traditional or revisited dishes.

To design high quality food experiences in relation to taste, nutritional value, and aesthetics, it was essential to recruit competent staff for the kitchen. This was done through collaboration with nationally and internationally famous vegan chefs who acted as external consultants.

The recruitment of competent waiters was also essential, especially for service quality. Personal contact is considered of paramount importance, especially in relation to guiding customers in their choice of products they might not be familiar with; for example, the seven varieties of plant-based milk offered at the café.

The owners observed that loyal customers tend to take the role of "guides" when they bring friends to the café. They also tend to engage in conversations with the waiters about ingredients and recipes. This suggests that a sort of cocreation occurs, with the café becoming a platform where, occasionally, the role of customer and supplier becomes blurred. With regard to this aspect, contacts with and among the customers and possible social interactions are facilitated through the arrangement of cultural events, such as concerts and book presentations, and, more recently, through the inclusion in the café of a small market with selected products.

The design process of the café was conducted using the services of a professional studio. The architects designed a space characterized by brightness, with a high ceiling and big windows, and elegance, with a combination of various materials, some designer furniture and pieces of art. Exclusively vegan and environmentally friendly materials were preferred to standard ones. This confirms those scholarly contributions about vegetarian restaurants showing concern about the environment.

With regard to the experience design principle concerning memorabilia, *Essenza* has developed some cooler-bags used during the summer so that customers can take their lunch on the beach and is now considering the design of T-shirts for the coming summer season.

10.7 Conclusions and practical Implications

This chapter has presented the case study of an Italian plant-based café with the aim to explore possible experience design strategies that plant-based restaurants might apply to target a market segment broader than one comprised of only vegans.

The findings suggest that the concept of neotribes, indicating fluid social groupings typical of postmodern consumerism, is suitable to capture the complexity of the vegetarian segment in those contexts where vegetarianism is a new and fast-growing phenomenon. At the same time, the case study suggests that there might be other relevant aspects that, in addition to or instead of the adherence to vegetarianism, should be taken into consideration when segmenting the market and developing experience design strategies. Gender might be one of these. Local culture, especially in relation to how vegetarianism/veganism is perceived, has emerged as a particularly important aspect.

The findings suggest that many of the aspects identified in the literature about food experiences can be applied to the case of plant-based restaurants. What is peculiar is the importance to choose a theme that is not perceived as being in conflict with the local culture but supports benefits and interests that are transversal to the various neotribes. Here, the nutritional aspect of food, the elegant and environmentally friendly aspect of the restaurant might be important aspects to consider. The design strategies that are dependent on qualified staff working in the kitchen and in contact with customers might be particularly challenging. Investing resources in the recruitment and training of staff is critically important to the quality of the food and service. This might be crucial to gain an early competitive advantage within the sector, characterized by a considerable growth of potential customers, and relatively limited knowledge and competence of practitioners.

References

Andersson, T. D., & Mossberg, L. (2004). The dining experience: Do restaurants satisfy customer needs? *Food Service Technology*, *4*(4), 171–177.

Bertella, G. (2018a). Vegetarian for a day or two. In A. Hardy, A. Bennett, & B. Robards (Eds.), *Neo-tribes: Consumption, leisure and tourism* (pp. 13–50). Cham: Palgrave McMillan.

Bertella, G. (2018b). Vegetarian ecofeminism: Emerging tourism practices by institutional entrepreneurs. In B. Grimwood, K. Caton, & L. Cooke (Eds.), *New moral natures in tourism* (pp. 86–98). Oxon: Routledge.

Bertella, G. (2018c). Animals off the menu: How animals enter the vegan food experience. In C. Kline (Ed.), *Animals, food, and tourism* (pp. 67–81). Oxon: Routledge.

Boswijk, A., Thijssen, T., & Peelen, E. (2006). *A new perspective on the experience economy. Meaningful experiences.* Amsterdam: European Centre for the Experience Economy. Pearson Education.

Cagnina, M. R., Cicero, L., &Osti, L. (2017). Dining motivations, experiences and loyalty at vegan/vegetarian restaurants. *SIM.* <https://bia.unibz.it/handle/10863/4881> Accessed 31.01.19.

Carù, A., & Cova, B. (2003). Revisiting consumption experience. A more humble but complete view of the concept. *Marketing Theory*, *3*(2), 267–286.

Cheng, C. H., Lin, S. Y., & Tsai, C. C. (2014). Investigating consumer preferences in choosing vegetarian restaurants using conjoint analysis. *Current Urban Studies*, *2*(03), 279–290.

Cole, M. (2008). Asceticism and hedonism in research discourses of veganism. *British Food Journal*, *110*(7), 706–716.

Cova, B., & Cova, V. (2001). Tribal aspects of postmodern consumption research: The case of French in-line roller skaters. *Journal of Consumer Behaviour*, *1*(1), 67–76.

Euripes. (2018). Euripes, Rapporto Italia 2018. Vegani e vegetariani sono il 7% della popolazione dai 18 anni in su. <http://eurispes.eu/news/eurispes-rapporto-italia-2018-vegani-e-vegetariani-sono-il-7-della-popolazione-dai-18-anni-in-su/> Accessed 04.04.19.

Flyvbjerg, B. (2006). Five misunderstandings about case-study research. *Qualitative inquiry*, *12*(2), 219–245.

Fox, N., & Ward, K. (2008). Health, ethics and environment: A qualitative study of vegetarian motivation. *Appetite*, *50*, 422–429.

Gustafsson, I. B., Öström, Å., Johansson, J., & Mossberg, L. (2006). The five aspects meal model: A tool for developing meal services in restaurants. *Journal of foodservice*, *17*(2), 84–93.

Ha, J., & Jang, S. (2010). Effects of service quality and food quality: The moderating role of atmospherics in an ethnic restaurant segment. *International Journal of Hospitality Management*, *29*(3), 520–529.

Hardy, A., & Robards, B. (2015). The ties that bind: Exploring the relevance neotribal theory to tourism. *Tourism Analysis*, *20*(4), 443–454.

ISTAT. (2018). *Popolazione Grosseto.* <https://www.tuttitalia.it/toscana/77-grosseto/statistiche/popolazione-andamento-demografico/> Accessed 04.04.19.

Janda, S., & Trocchia, P. J. (2001). Vegetarianism: Toward a greater understanding. *Psychology & Marketing*, *18*(12), 1205–1240.

Jang, S., & Namkung, Y. (2009). Perceived quality, emotions, and behavioral intentions: Application of an extended Mehrabian–Russell model to restaurants. *Journal of Business Research*, *62*(4), 451–460.

Knutson, B. J., & Patton, M. E. (1993). Restaurants can find gold among silver hair. *Journal of Hospitality and Leisure Marketing*, *1*, 79–90.

Lea, E. J., Crawford, D., & Worsley, A. (2006). Consumers' readiness to eat a plant-based diet. *European Journal of Clinical Nutrition*, *60*(3), 342–351.

Leahy, E., Lyons, S., & Tol, R. S. (2010). An estimate of the number of vegetarians in the world (no. 340). *ESRI working paper.* <https://www.econstor.eu/handle/10419/50160> Accessed 28.05.20.

Lee, S. J., & Crange, D. A. (2007). The relative importance of menu attributes at point of menu selection through conjoint analysis: Focused on adolescents. *Journal of Foodservice Business Research*, *10*, 3–18.

Mayring, P. (2000). Qualitative content analysis. *Forum: Qualitative Social Research*, *1*(2). Available from http://www.qualitative-research.net/index.php/fqs/article/view/1089/2385, Accessed 17.09.19.

Pine, B. J., & Gilmore, J. H. (1998). Welcome to the experience economy. *Harvard Business Review*, *76*, 97–105.

Pine, B. J., & Gilmore, J. H. (2013). The experience economy: Past, present and future. In J. Sundbo, & F. Sørensen (Eds.), *Handbook of the experience economy* (pp. 21–44). Cheltenham: Edward Elgar.

Rivera, M., & Shani, A. (2013). Attitudes and orientation toward vegetarian food in the restaurant industry: An operator's perspective. *International Journal of Contemporary Hospitality Management*, *25*(7), 1049–1065.

Rosenfeld, D. L. (2018). The psychology of vegetarianism: Recent advances and future directions. *Appetite*, *131*, 125–138.

Rothgerber, H. (2014). A comparison of attitudes toward meat and animals among strict and semi-vegetarians. *Appetite*, *72*, 98–105.

Ruby, M. B. (2012). Vegetarianism. A blossoming field of study. *Appetite*, *58*, 141–150.

Ryu, K., & Jang, S. (2008). DINESCAPE: A scale for customers' perception of dining environments. *Journal of Foodservice Business Research*, *11*(1), 2–22.

Shani, A. (2010). Vegetarianism. In A. Pizam (Ed.), *International encyclopaedia of hospitality management* (pp. 674–676). Oxford: Butterworth-Heinemann.

Shani, A., & DiPietro, R. B. (2007). Vegetarians: A typology for foodservice menu development. *Hospitality Review*, *25*(2), 65–73.

Sobal, J., Bisogni, C. A., Devine, C. M., & Jastran, M. (2006). A conceptual model of the food choice process over the life course. *Frontiers in Nutritional Science*, *3*, 1–18.

Soriano, D. R. (2002). Customers' expectations factors in restaurants: The situation in Spain. *International Journal of Quality and Reliability Management*, *19*, 1055–1068.

Tardivo, G., Thrassou, A., Viassone, M., & Serravalle, F. (2017). Value co-creation in the beverage and food industry. *British Food Journal*, *119*(11), 2359–2372.

Van der Kooi, M. (2010). The inconsistent vegetarian. *Society and Animals*, *18*, 291–305.

Walter, U., Edvardsson, B., & Öström, Å. (2010). Drivers of customers' service experiences: A study in the restaurant industry. *Managing Service Quality: An International Journal*, *20*(3), 236–258.

Event catering: enhancing customer satisfaction by creating memorable holistic food experiences

11

Susanne Doppler, Adrienne Steffen and Lisa-Marie Wurzer
University of Applied Sciences Fresenius Heidelberg, Heidelberg, Germany

11.1 Introduction

Catering is commonly associated with social events at which food and beverages are served in a professional manner, since a successful event is often combined with the consumption of food and drinks (Sharples, 2014, p. 167; Shaw & Ferdinand, 2012, p. 253).

New technologies as well as the increase in travel and tourism have increased the popularity of the catering business (Mhlanga, 2015). According to the corporate travel group Carlson Wagonlit Travel, meetings, incentives, conferences, and exhibition markets, collectively known as MICE, which is mainly catered for by hotels, restaurants, and contract caterers, are worth an estimated US$650 billion (Shaw & Ferdinand, 2012, p. 254). Additionally according to the IBISWorld UK market research report of the catering service industry (2018), "corporate events account for approximately half of total revenue in the industry [in the UK]."

Therefore it is necessary to understand which factors influence a customer experience (Kauppinen-Räisänen, Gummerus, & Lehtola, 2013). An extensive review of food experience dimensions in restaurants, such as food quality, product quality, service quality, and atmosphere/physical facilities, is provided by Kleinhans and Van Heerden (2016). Based on their review, the authors concluded that dining experience dimensions are unique to the setting and circumstances of the type of the dining place. However, the number of academic studies representing the strategic relevance and importance of food to the event experience is low (Sharples, 2014, p. 168). Therefore the purpose of this study is to fill the gap in this specific sector, given the abovementioned increased importance of food services in the event industry today. The aim of the study is to answer the question: how can an event caterer create memorable and holistic food experience to improve customer satisfaction?

Case Studies on Food Experiences in Marketing, Retail, and Events. DOI: https://doi.org/10.1016/B978-0-12-817792-1.00011-3

11.2 Literature review

11.2.1 The purpose of event catering

Events are an excellent communication platform which contains all facets of communications like no other marketing instrument (Doppler, 2016, p. 22). Event catering is one of these facets of communication. Therefore the food component of an event should be "part of the master-plan and eventscape" (Sharples, 2014, p. 169) thus reflect the strategic objectives of the event. Food experiences at events allow for the expression of prestige, status (Sharples, 2014, p. 171), and hospitality (Sharples, 2014, p. 176). Food offerings at events acknowledge the need of refreshment during events and support the performance of participants and therefore long-term health benefits (Sharples, 2014, p. 170). Furthermore event catering offers a platform to interact, network and communicate in a casual and informal atmosphere (Sharples, 2014, p. 173). The quality of the catering at an event has an impact on its success, what means that an event's success can rise or fall depending on the catering (ACS, n.d.; Sharples, 2014, p. 167). Additionally catering has identified itself as more than just a matter of serving delicious food but also has evolved to become an artisanal affair. Caterers are now creating an experience that involves all the senses (Thinius & Untiedt, 2013, p. V).

11.2.2 Customer food experience

An experience that involves all senses is largely defined by customer experience (for further details on customer experience see Steffen & Doppler, 2020). With regard to these experiences, the customer journey has to be taken into consideration (Ferdinand & Kitchin, 2017, p. 272).

Ideally the customer journey goes through a targeted dramaturgy. Dramaturgy relies on the ability to tell a story enticingly and compellingly, to create tensions, generate attention, and avoid boredom, for example, through interaction between a customer and a product or a company, which causes the customer to react (Bruhn, 2016, p. 78; Doppler, 2016, p. 18) and cocreate values for both the customer and the company. The aim of customer experience management is to create positive customer experiences and a holistic experience at all touchpoints along the entire customer journey, to generate customer enthusiasm, and to sustainably bind customers to the company (Ferdinand & Kitchin, 2017, p. 272).

The purpose of eating out has changed over the years and consumers tend to dine out to celebrate special occasions with others rather than simply satisfying their basic hunger needs, partly to distract themselves from their daily life by remembering such experience (Abdelhamied, 2011). Santos (2002) states that dining experience rely on the combination of tangible and intangible dimensions a customer experiences when eating out. On the one hand, the customer makes a decision with respect to food and quality, but on the other hand expects a complete dining experience, which is influenced by various factors (Kleinhans & Van Heerden, 2016). This "[...] dining experience is [...] the core concept in the catering industry" (Kleinhans & Van Heerden, 2016, p. 2), and customer experience has been identified as the key differentiator (Golani, 2017). The study of Kauppinen-Räisänen et al. (2013) shows that food experiences are mainly created through the sensory, emotional and social dimension.

To understand and create a holistic food experiences for customers, it is necessary to identify which factors contribute to enhance customer satisfaction (Tanasca, Purcarea, & Popa, 2014).

11.2.3 Customer food experiences as a driver for customer satisfaction

The dining experience (Fraikue, 2014) and customer experience (Caruana, 2002 as cited in Klaus & Maklan, 2013, p. 229; Shankar, Smith, & Rangaswamy, 2003) are key determinants of customer satisfaction. Researchers suggest that experiences drive satisfaction (Arora, 2012; Kleinhans & Van Heerden, 2016; Sancheti, Karakoti, & Dixit, 2018; Shankar et al., 2003). Therefore when trying to create memorable, holistic food experiences, customer satisfaction has to be taken into account (Koththagoda & Weerasiri, 2017). Furthermore customer satisfaction influences customer revisit intentions (He & Song, 2009). In addition, Saad Andaleeb and Conway (2006, p. 3) state that "customer satisfaction is at the heart of marketing." In other words, satisfaction is viewed as the main principle of marketing strategy (Arora, 2012, p. 353). Harr (2008) developed different so-called service dimensions which have an impact on satisfaction. The most important ones are assurance, empathy and tangibles which also contribute to a high level of customer satisfaction. Research indicates that there is a relationship between customer satisfaction and service quality (Golani, 2017; He & Song, 2009; Koththagoda & Weerasiri, 2017). The analysis of Koththagoda and Weerasiri (2017) shows that physical environment, service, and quality are the core determinants in generating customer satisfaction. However, others came to the conclusion that quality is not equivalent to satisfaction, but related (Parasuraman, Valarie, & Leonard, 1988). This assumption can be challenged by the result of Golani's study (2017). It expresses that food quality is of decisive importance as it contributes to customer satisfaction.

11.2.4 Factors influencing memorable, holistic food experiences at an event

In the modern hypercompetitive environment, companies are increasingly drawn to the idea of creating memories instead of good settings to generate a higher economic value (Tanasca et al., 2014). Therefore factors that create a memorable, holistic dining experience are examined (see Fig. 11.1).

The core principle of organizing an event is to create memorability as well as succeeding at telling a story (Thinius & Untiedt, 2013), as the perception in the

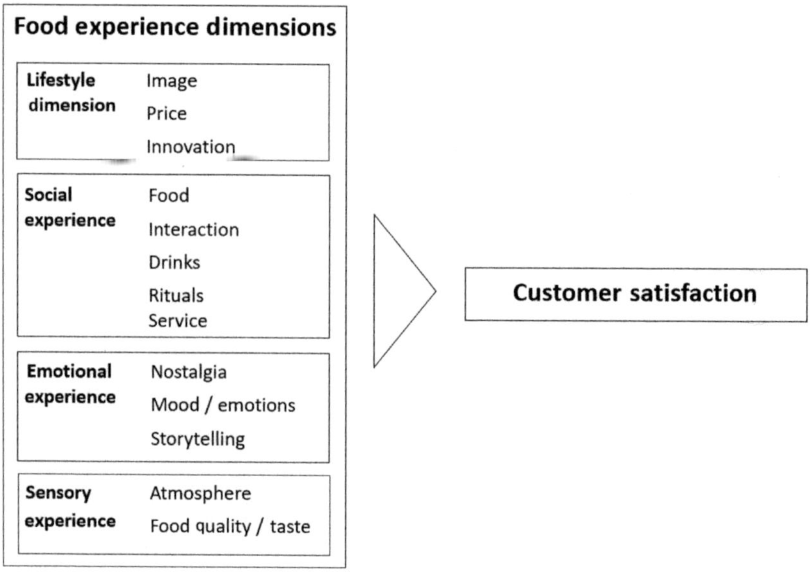

Figure 11.1 Conceptual framework, own illustration.

consumer's mind is of decisive importance to his or her evaluation of the event (Sancheti et al., 2018).

The event catering industry relies on the core concept of dining experiences as there are several factors or dimensions which can influence the dining experience (Kleinhans & Van Heerden, 2016). Their study shows that food experiences depend on service, quality, atmosphere, food, physical facilities, price, health, cleanliness, and social dimensions (Kleinhans & Van Heerden, 2016). The study of Hansen, Jensen, and Gustafsson (2005), support this view and show that restaurant experiences are defined by five core aspects: the core product (food), the restaurant interior, the personal social meeting (customer and expert), the company the customer were in, and the restaurant's atmosphere. With regard to the atmosphere, the study of Ha and Jang (2012) emphasize the impact of atmosphere on a customer's perceived quality. Highlighting the important role of the atmosphere for quality perception in creating positive behavioral intentions, it suggest that the atmosphere could improve or diminish the perceived quality of an event by customers.

In addition, Mhlanga (2015, p. 1) states that customer satisfaction is influenced by the impression of service quality, product quality, and price as well as situational factors and personal factors. This can be amplified by the SERVQUAL study of Parasuraman et al. (1988).

Moreover, McGaugh (2013, p. 10402) states that the "strength of memories of events varies with the emotional significance of the events." Furthermore Phillips and Baumgartner (2002) explain that positive consumption emotions will have a positive impact on satisfaction. According to McGaugh (2013, p. 10402) emotional excitement improves the storage of memories. In other words, emotional arousal is strongly linked to memory. Moreover, Jin, Line, and Merkebu (2015) and He and Song (2009) state that creating postconsumption emotions is positively linked to revisiting intentions.

Additionally rituals such as mortuary food also contribute to a memorable experience (Holtzman, 2006, p. 372).

11.3 Methodology

The study explores the conceptual framework (Fig. 11.1). To explore the conceptual framework, a qualitative research approach was chosen (Döring & Bortz, 2016, pp. 182–193; Thomas, 2016, p. 120). Primary and secondary data were collected in the form of a case study approach. The catering company Feinkost Käfer (hereinafter referred to as Käfer) was selected as the subject of the case study as it is one of the world's leading event catering companies (Statista, 2017). A case study approach was chosen to investigate how Käfer generates customer satisfaction through creating customer food experiences in the event catering sector. The case presented herein is a single, retrospective case using a qualitative research approach (Döring & Bortz, 2016, pp. 182–192; Thomas, 2016, p. 120). Primary data were collected in form of two semistructured interviews with Käfer business customers and four semistructured expert interviews with Käfer Experts. The interviewees were selected by a gatekeeper at Käfer. The semistructured interviews helped to emphasize the factors and influences depicted in the conceptual framework. All interviews were transcribed, the content was analyzed with an inductive coding approach of Mayring as cited in Döring and Bortz (2016, p. 376). Relevant text passages were translated from German to English and paraphrased. Codes were established and reviewed again after a reduction of the paraphrased sentences. Secondary data were obtained from the company's webpage to explore Feinkost Käfer's values.

11.3.1 The case Feinkost Käfer

Feinkost Käfer GmbH was founded in 1930 by Paul and Elsa Käfer in Munich, Germany where they established their first business in form of a grocery store. In 1933 the grocery store moved to the present head office. In 1959 Gerd Käfer added the party service segment to the company (Käfer, n.d.a). The Käfer party service is one of the world's leading event caterers (Statista, 2017). As a full-service provider with an international network, the party service organizes an average of 1250 events worldwide with approximately 220,000 guests per year. Over the past years, the company has opened 18 restaurants and bistros around the world. Käfer is the European market leader in the premium segment of party services (Käfer, n.d.a). In 2017 1500 Experts generated revenue of €136.3 m (Käfer, n.d.b).

11.4 Results

In a first step, Käfer's values, on which the company's success is based, are explored. In the second step, the factors for a memorable holistic food experience are examined from the point of view of the supplier (expert) and the customer.

11.4.1 *Feinkost Käfer's five values*

The Käfer Group has four business branches: party service, gastronomy, retailing, and licensing. These economic components are strongly combined with the guiding principle of the company "Passion for Quality" which happens to be the key determinant of the company's success (Käfer, n.d.c). The quality of their products is characterized by the passion in which the experts carry out their work. Therefore five individual values are guiding their action. These values define the demands Käfer places on themselves, their products, services, partners and experts (Käfer, n.d.c). The values "lovable, unique, unexpectantly, exquisite and emotional" are guidelines for Käfer's actions. Moreover, naturalness and authenticity are an important factor which Käfer considers in their appearance (Käfer, n.d.c).

First, as being lovable and credible is a challenging task, Käfer secures the trust of their customers through the accessibility and personal commitment. Second, when talking about uniqueness, Käfer's experts combine the perfection and exclusivity of selected products in a unique range and a perfect presentation with a lively lightness that offers their customers maximum enjoyment and thus always a special joy of life (Käfer, n.d.c). Third, offering the unexpected is Käfer's permanent goal by offering special goods and perfect service, as well as gastronomic highlights and individual, unforgettable events. The shopping experience at Käfer also offers positive surprise time which is emphasized by the personal service and the charming presentation in the grocery store. Moreover, experts are invited to special workshops to gain knowledge in the field of their products to outstand in front of customers and to express their guiding principle, passion for quality (Käfer, n.d.c). Fourth, when talking about exquisite, Käfer strives for the constant verification and adherence of existing standards. Furthermore they expect their supplier to only produce the highest quality as this is the only way to create the highest enjoyment. Fifth, at Käfer, being emotional means living the passion for the highest level of enjoyment. Every product bearing the Käfer name has a personal connection to the family business. This value is one key aspect as emotionality is more relevant today than ever (Käfer, n.d.c).

11.4.2 *Creating customer food experience*

11.4.2.1 *Spontaneous reaction to the term customer food experience*

When asking the interviewee for their spontaneous reaction to the term "customer food experience" several themes were mentioned. These included the way of how the food is cooked, presented and served as well. Besides experiences with designed food there seems to be a return to original taste experiences. Furthermore a potential can be identified that can be traced back to the interaction between customer and supplier. While the customer wants to be surprised, the "cook" also tells about himself. This creates the potential for a common narrative. On a more abstract level, the combination of tradition and innovation appeared to be relevant when it comes to the creation of customer food experience.

> *Creating customer food experience is a combination. On the one hand, I think of for instance the molecular cuisine which is focused on experimental dishes, which surprises the customer. On the other hand, I think the quality of the products is of decisive importance, which means to present something raw and untreated in order to create this special traditional taste experience. Moreover, it is important to know and to understand the customer in order to generate this individuality for him which generates this surprising effect. (Expert 2, paraphrased)*

11.4.2.2 Factors influencing customer food experience

In this section, the factors which create a memorable holistic food experience to explore customer satisfaction are analyzed. All factors outlined in Fig. 11.1 were mentioned except nostalgia and rituals. In general, the main factors mentioned were quality, service and atmosphere, whereas the food should be combined with the atmosphere and ambience of the event design and location.

> *It is rare that only the food remains in the memory, mostly the whole event stays in memory and of course service is part of it. Furthermore, the ambience, the whole atmosphere and of course the food are linked to it. (Expert 1)*

Concerning quality, it is important that the caterer offers the best quality of products available on the market which means, constantly looking for high-performance suppliers offering high-quality products and ingredients:

> *It is about which supplier can offer the best product or ingredient. Therefore, I need a balance of different suppliers. (Expert 2)*

Customers also point out that the unobtrusive service has a strong impact on customer food experience.

> *Yes, this silentness. That we discuss this and it will be implemented quietly and uncomplicated and exactly the same will be delivered. It's like those brownies and fairies, who do everything in the background so that you notice something happens to your complete satisfaction and you don't have to worry about it. (Customer 2)*

Furthermore uniqueness and innovations in preparing and presenting food is a key success factor to caterers.

> *[...] innovation and constantly new product development. For example, we made trays for a hot stone, we produced them ourselves. Therefore, only we have them,*

no other caterer has them. Or we have had special barbecues [. . .] that are unique and can't be bought anywhere else. I believe that you have to constantly develop product innovation that others just can't offer. (Expert 4)

But at the same time, customers emphasize the relevance of the price.

The most important factors are price, quality and product variety. (Customer 2, paraphrased)

In the field of food presentation, the activation of different senses as well as customer interaction seems to play a central role to create food experiences

It is important to reach all the senses. Furthermore, interaction with the customer is important. For instance, cooking in front of the customers, as this appeals to what a customer sees and smells. Moreover, show cooking underlines the idea of interaction. (Expert 4, paraphrased)

Concerning the creation of customer food experiences, emotions have to be considered. However, in some way it is sometimes difficult for caterers to arouse emotions. Therefore it is necessary to differentiate between a business occasion and a private occasion:

It depends on the occasion. If it is a business meeting, emotions do not play an important role. If it is a special occasion, emotions play an important role. Therefore, we try to offer noble instead of basic staff. (Expert 1)

To arouse emotions, appealing to all the senses and surprising the customer by creating something unexpected by offering special food ingredients, smell and taste as well as an unexpected food presentation are the keys to evoke emotions.

Storytelling is used as a tool to be remembered. In event catering, the focus should lie on the experience of the story and less on background information like the origin and production of the food, as is more common in the retail trade. Therefore event caterers implement a narrative by adapting the food to the matching theme of the event.

Storytelling isn't quite as strong at events yet as with products in the delicatessen shop, because I can of course also describe the products there. I have package inserts everywhere as well as written inserts, where I can read them. (Expert 2)

The birthday of [anonymized name] was memorable as they tried to tell a story. Actually, his story in the form of different food stations and different acts. This caused an experience which is unforgettable. (Customer 1)

11.4.3 Event catering and customer satisfaction

In terms of business events, event catering supports and underlines for instance the product or brand marketed by the client. This aspect seems to be crucial for business customer catering, as the emotional arousal and satisfaction has an impact even on buying decisions:

> *Yes, it underlines the customer, so if a business customer wants to present a product, then we will underline the quality of the product with our catering. (Expert 2)*
>
> *I think [food experience] offers satisfaction in sum. Everyone likes to eat. The old German saying 'the way through a man's heart is through his stomach' also goes through the customer in some way. Therefore customer love goes through the stomach and if the customer has felt well in my shop because of an event catering which is supposed to generate such feeling, then I think to myself that he says okay, I felt good now at [anonymised brand] and therefore before I buy another car now, I first think about [anonymised brand].(Customer 1)*

Moreover, if customers are satisfied it also creates a positive word-of-mouth and increases customers through recommendations:

> *This is also a topic, you create moments of surprise and everyone talks about it. (Expert 4)*
>
> *The fact that if it works out well, I do it again and again and then I recommend it. (Customer 2)*

11.5 Discussion

It was demonstrated that there are several factors that influence the creation of customer food experiences in general to enhance customer satisfaction. The most frequently mentioned ones are (1) the quality of food, smell, and taste, (2) the uniqueness of presentation, (3) the service itself by means of being empathetic and attentive to the needs of the customer in the preevent communication, as well as offering a silent and unobtrusive service during the event, and (4) the fit between event design, atmosphere, and food, which is in line with the study of Kleinhans and Van Heerden (2016). This result shows that the sensory experience dimension is of the greatest importance when creating memorable holistic food experiences which clearly confirms the findings of Kauppinen-Räisänen et al. (2013).

The case of Käfer shows that the most important factor to enhance the overall satisfaction is to surprise the customer with something unexpected. In addition, Käfer always refers back to their five individual values which are the guidelines towards their action. Rather than letting customers create their own values towards a food experience as it is described by Johns (2014), Käfer aims to create these five values for their customers in advance, to gain satisfaction (Käfer, n.d.c).

Furthermore one factor which was constantly mentioned is the way how the food is presented as it influences customer memorability and therefore can have an effect on customer satisfaction. Moreover, it is necessary to interact and to prepare the food in front of the guests to make a food experience memorable and to enhance customer satisfaction (Bruhn, 2016, p. 78; Doppler, 2016, p. 18).

Offering food at an event can enhance customer satisfaction which emphasizes the study of Fraikue (2014), and thus, has a very strong impact on repurchasing decision and revisit intentions which confirms the findings of He and Song (2009).

Nostalgia and rituals were not mentioned once by any interviewee. One possible explanation is that participants answered spontaneously and did not have time to think about these factors thoroughly. However, the literature on nostalgia and rituals claim that these factors are crucial to create a memorable holistic food experience (Batsell, Brown, Ansfield, & Paschas, 2002; Holtzman, 2006; Kauppinen-Räisänen et al., 2013). Therefore these factors should not be neglected.

Compared to the study of Jin et al. (2015), menu innovation is a key way to gain memorability. The case study shows that the combination of innovation with tradition creates an overall satisfaction for their customers and influences memorability.

11.6 Managerial implications

As a result of the study framework it is recommended to consider several aspects in strategic and operational catering planning as shown in Fig. 11.2. As a starting point for all activities, event caterers should define and agree on individual values about oneself, the products, service, partners and experts, and deliver them to the customer in advance. These values serve as a guideline for every customer contact and will be recognized by the customers, which, as a consequence, enhances customer satisfaction.

Event catering must be integrated into the conception of events from the very beginning and should no longer be seen as a matter of course, even for smaller events. To meet the long-term quality requirements of customers a quality management system should come into consideration, which covers product and service quality to ensure that suppliers as well as experts deliver and act on a high-performance level during the pre- and live event phase. Based on individual values, all relevant areas in the customer and supplier management, the conception, creation and presentation of food, including the supply chains should be reflected.

11.7 Conclusions

The case has shown that event catering plays a central role for the quality of experience and customer satisfaction. One essential prerequisite for creating memorable, holistic food experiences is that the caterer defines individual values as a guideline for their service. When it comes to the implementation, the quality of the food and ingredients, the service quality in the preevent and live event phase, the presentation of the food, and the harmony between event design, atmosphere, and food are decisive factors for a memorable holistic food experience. Even particularly unexpected elements in the experience create unforgettable moments and contribute to

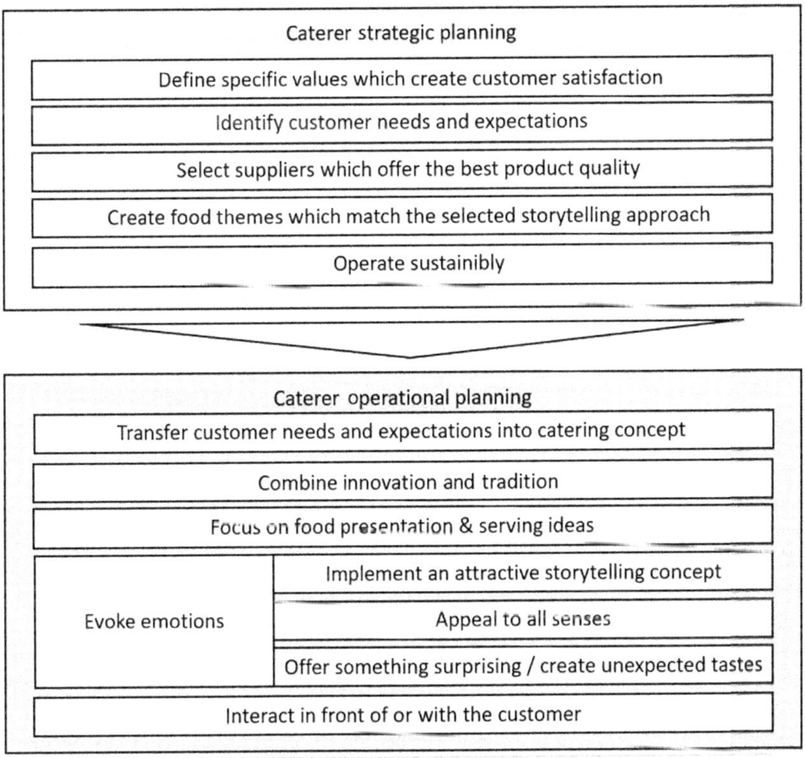

Figure 11.2 Aspects to consider in strategic and operational catering planning.

an increase in customer satisfaction. The interviews showed that the sensory dimension seems to be of utmost importance in this context.

References

Abdelhamied, H. H. S. (2011). Customers' perceptions of floating restaurants in Egypt'. *Anatolia: An International Journal of Tourism and Hospitality Research*, *22*(1), 1–15.

ACS Distance School. (n.d.). Catering at events: Determining food and beverages needs. <https://www.acsedu.co.uk/Info/Hospitality-and-Tourism/Hospitality/Catering-at-Events.aspx> Accessed 2002.19.

Arora, R. (2012). A mixed method approach to understanding the role of emotions and sensual delight in dining experience. *Journal of Consumer Marketing*, *29*(5), 333–343.

Batsell, W. R., Brown, A., Ansfield, M., & Paschas, G. (2002). You will eat all of that: A retrospective analysis of forced consumption episodes. *Appetite*, *38*, 211–219.

Bruhn, M. (2016). *Qualitätsmanagement für Dienstleistung: Handbuch für ein erfolgreiches Qualitätsmanagement Grundlagen - Konzepte - Methoden* (Vol. 10). Auflage, Wiesbaden: Springer Gabler.

Doppler, S. (2016). *B2B-event-marketing*. Konstanz: UVK Vertragsgesellschaft.

Döring, N., & Bortz, J. (2016). *Forschungsmethoden und Evaluation: in den Sozial- und Humanwissenschaften* (5th ed.). Heidelberg: Springer.

Ferdinand, N., & Kitchin, P. J. (2017). *Events management. An international approach* (2nd ed.). London: Sage.

Fraikue, F. B. (2014). *The meal experience of customers of grade 3 restaurants in Sekondi-Takoradi*. Ghana: Takoradi Technical University.

Golani, N. (2017). Factors influencing customer satisfaction & customer delight in fine dining restaurants. *International Journal for Research in Applied Science and Engineering Technology*, *V*(II), 629–638.

Ha, J., & Jang, S. C. (2012). The effects of dining atmospherics on behavioral intentions through quality perception. *Journal of Services Marketing*, *26*(3), 204–215.

Hansen, K. V., Jensen, Ø., & Gustafsson, I. B. (2005). The meal experiences of á la carte restaurant customers. *Scandinavian Journal of Hospitality and Tourism*, *5*(2), 135–151.

Harr, K. K. L. (2008). *Service dimensions of service quality impacting customer satisfaction of fine dining restaurants in Singapore* (thesis). University of Nevada, Reno, NV.

He, Y., & Song, H. (2009). A mediation model of tourists' repurchase intentions for packaged tour services. *Journal of Travel Research*, *47*(3), 317–331.

Holtzman, J. D. (2006). Food and memory. *Annual Review of Anthropology*, *35*, 361–378.

IBISWorld. (2018). *Catering services: UK market research report*. <https://www.ibisworld.co.uk/industry-trends/market-research-reports/accommodation-food-service-activities/catering-services.html> Accessed 21.02.19.

Jin, N. P., Line, N. D., & Merkebu, J. (2015). Examining the impact of consumer innovativeness and innovative restaurant image in Upscale Restaurants. *Cornell Hospitality Quarterly*, *57*(3), 268–281.

Johns, N. (2014). Consumer behaviour in the food service industry: A review consumer behaviour in the food service industry: A review. *International Journal of Hospitality Management*, *21*, 119–134. (September).

Käfer. (n.d.a). *Firmenportrait*. <https://static.feinkost-kaefer.de/media/pdf/86/fc/dd/Firmenportrait_20185b34f3b78ba42.pdf> Accessed 02.04.19.

Käfer. (n.d.b). *Käfer in Zahlen*. <https://www.feinkost-kaefer.de/inzahlen> Accessed 02.04.19.

Käfer. (n.d.c). *Unsere Werte*. <https://www.feinkost-kaefer.de/unserewerte> Accessed 02.04.19.

Kauppinen-Räisänen, H., Gummerus, J., & Lehtola, K. (2013). Remembered eating experiences described by the self, place, food, context and time. *British Food Journal*, *115*(5), 666–685.

Klaus, P., & Maklan, S. (2013). Towards a better measure of customer experience. *International Journal of Market Research*, *55*(2), 227–246.

Kleinhans, E., & Van Heerden, C. (2016). A review of dining experience dimensions over two decades. *African Journal of Hospitality, Tourism and Leisure*, *5*(4).

Koththagoda, K., & Weerasiri, S. (2017, January). Factors influencing dining experience on customer satisfaction and customer loyalty: with special reference to Restaurant Industry in Star Class Hotels in Sri Lanka.

McGaugh, J. L. (2013). Making lasting memories: Remembering the significant. *Proceedings of the National Academy of Sciences*, *110*, 10402–10407.

Mhlanga, O. (2015). Electronic meal experience: A gap analysis of online Cape Town restaurant comments. *African Journal of Hospitality, Tourism and Leisure*, *4*(1).

Parasuraman, A., Valarie A.Z. and Leonard L.B. (1988). Servqual: A multiple-item scale for meassuring consumer perceptions of service quality. *Journal of Retailing, 64(1)*.

Phillips, D. M., & Baumgartner, H. (2002). The role of consumption emotions in the satisfaction response. *Journal of Consumer Psychology*, *12*(3), 243–252.

Saad Andaleeb, S., & Conway, C. (2006). Customer satisfaction in the restaurant industry: An examination of the transaction-specific model. *Journal of Services Marketing*, *20*(1), 3–11.

Sancheti, K., Karakoti, R., & Dixit, A. (2018). Variables contributing to the time spent by customers in the dining experience. *International Journal of Academic Research and Development*, 853–855.

Santos, J. (2002). From intangibility to tangibility on service quality perceptions: A comparison study between customers and service providers in four service industries. *Managing Service Quality*, *12*(5), 292–302.

Shankar, V., Smith, A. K., & Rangaswamy, A. (2003). Customer satisfaction and loyalty in online and offline environments. *International Journal of Research in Marketing*, *20*(2), 153–175.

Sharples, L. (2014). Food and function. In L. Sharples, P. Crowther, D. May, & C. Orefice (Eds.), *Strategic event creation*. Oxford: Goodfellow.

Shaw, S. J., & Ferdinand, N. (2012). *Events management – An international approach, business & economics*. London: Sage.

Statista. (2017). *Ranking of the leading system catering companies in Germany from 2016 to 2017, by revenue (in billion euros)*. <https://www.statista.com/statistics/561002/system-catering-companies-germany/> Accessed 02.04.19.

Steffen, A., & Doppler, S. (2020). Introduction to the experience economy. In S. Doppler, & A. Steffen (Eds.), *Case studies on food experience: Insights for marketing, retail and events*. Woodland: Elsevier.

Tanasca., Purcarea., & Popa, V. (2014). Quality in service industry and customer experience management (CEM) case study: Restaurant industry. *Supply Chain Management Journal*, *5*(1), 77–96.

Thinius, J., & Untiedt, J. (2013). *Events - Erlebnismarketing für alle Sinne: Mit neuroyaler Markenkommunikation Lebensstile inszenieren* (2nd ed.). Wiesbaden: Springer Gabler.

Thomas, G. (2016). *How to do your case study* (2nd ed). London: Sage.

"We have our style!" Optimizing food experiences in personalized catering boutique concepts

12

Franklin Velasco-Vizcaíno[1], Attila Pohlmann[1] and Mauricio Cepeda[1,2]
[1]Department of Marketing, Business School, Universidad San Francisco de Quito, Quito, Ecuador, [2]Department of Hospitality Management, School of Hospitality Management and Culinary Arts, Universidad San Francisco de Quito, Quito, Ecuador

12.1 Introduction to catering boutique concepts

The thought of catering services often conjures up the image of a purely logistical challenge to provide food to a large number of people—often at remote sites. However, emerging luxury boutique catering services increasingly expand their capacities to provide holistic experiences, which go beyond the mere satiation of hunger.

In the competitive hospitality marketplace, boutique concepts provide unique differentiation possibilities. Boutique concepts are often characterized as small, entrepreneurial, innovative, and as tightly focused on the customer's experience (Jones, Day, & Quadri-Felitti, 2013; Kurgun, Bagiran, Ozeren, & Maral, 2011; Mcintosh & Siggs, 2005). Thus expanding innovative capacities and managing the experiential aspects of service are crucial for successful businesses. Especially boutique concepts can set themselves apart from standardized experiences by offering their customers uniqueness, quality, and highly personalized levels of service (Aggett, 2007; Victorino, Verma, Plaschka, & Dev, 2005). In many instances, the customers' participation in the service process (Mustak, Jaakkola, Halinen, & Kaartemo, 2016) provides countless value cocreation opportunities (Lusch & Vargo, 2006; Schembri, 2006; Vargo & Lusch, 2008).

In this study, we associate the boutique concept with the outstanding food experiences that boutique catering services provide. This study highlights prominent breakthrough differentiation attributes that Epikus, a successful boutique catering service in Ecuador, integrates in its core service orientation and service delivery process. Epikus' success story includes several examples of value cocreation opportunities (Vargo, Maglio, & Akaka, 2018), expansion of organizational boundary-spanning roles (Thompson, 1962), and service-dominant logic (Vargo & Lusch, 2008), all of which coexist within the boutique concept. For instance, Epikus demonstrates customer centricity, a holistic perspective for service delivery, intensification of service atmospherics (i.e., food presentation and event decoration), customer relationship building, customers—employees one-on-one interactions, and

Case Studies on Food Experiences in Marketing, Retail, and Events. DOI: https://doi.org/10.1016/B978-0-12-817792-1.00012-5

the magnification of elegance. Findings from this study suggest that these aspects are strongly related to the boutique concept and show a path to induce prosperous food experiences to catering services clientele.

To capitalize on these opportunities, catering service providers need to be equipped to manage its experiential components—the social, relational, and intersubjective aspects of the service experience (Helkkula, 2011; Pullman & Gross, 2004). However—posing some managerial challenges—these aspects often do not stand immutably on their own but are dynamically interpreted by customers depending on their individual personal and situational contexts (Gummesson, 2008; Prahalad & Ramaswamy, 2004; Vargo & Lusch, 2008). Such context-dependent conceptualizations have received considerable attention in hospitality and tourism research (e.g., Chathoth, Ungson, Harrington, & Chan, 2016; Shaw, Bailey, & Williams, 2011; Victorino et al., 2005) and provide the conceptual lens for the following case study.

12.2 Case study method

The goal of this research was to gain a deeper understanding of how managers and frontline employees of luxury catering boutiques provide innovative and experiential service offerings. Epikus Boutique Catering, a catering service located in Quito-Ecuador, has been operating for over 17 years as part of a luxury restaurant and culinary school, both of which are associated with Universidad San Francisco de Quito as well. The research approach for this study initially identified managers and staff who were willing to participate in personal interviews relating to critical incidents (Bott & Tourish, 2016; Cope & Watts, 2000; Flanagan, 1954), which provided them with both challenges and learning opportunities in the last 3 years. Fig. 12.1 shows the organizational chart of Epikus.

Epikus is a boutique catering service as defined by our interviewees and the *clientèle élégante* they serve. Epikus' growth and recognition has been steady since its foundation; evinced by its sales growth illustrated in Fig. 12.2. Epikus generates most of its revenue from wedding events (about 60%). Other social events that Epikus caters to include wedding anniversaries, graduation ceremonies, bar mitzvahs, first communion celebrations, birthday parties, and any social event that needs a touch of luxury and a *service cinq étoiles*. Appendixes A and B show photographs of the food presentation and service experience that Epikus delivers to its clients.

The critical incident technique (CIT, Byrne, 2001; Flanagan, 1954) is a qualitative approach that uses semistructured interviews to collect accounts of direct actions and behaviors from participants during a specific situation. The main objective of the CIT is to present an incident (e.g., factual events that occur during the management of a boutique catering service) to the interviewee and collect insights on how the interviewee reacted to the incident (e.g., stories of customers that were successfully served, accounts of the customer decision making processes, accounts

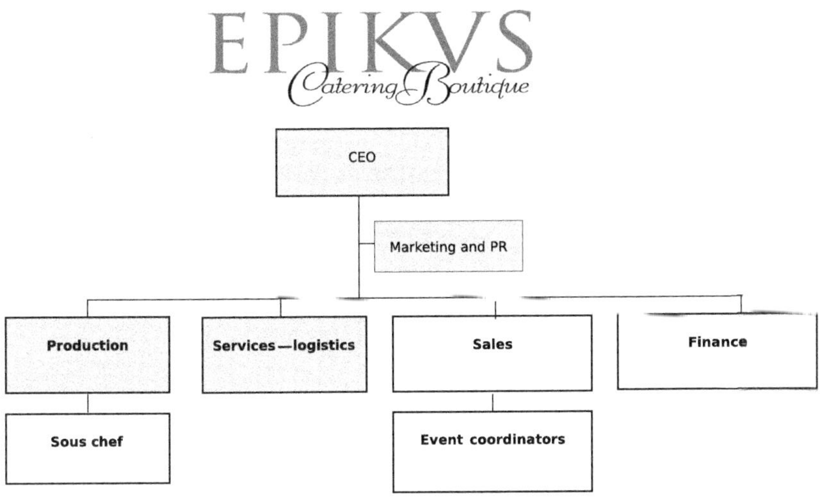

Figure 12.1 Organizational chart Epikus. Shading indicates departments that participated in the interviews.

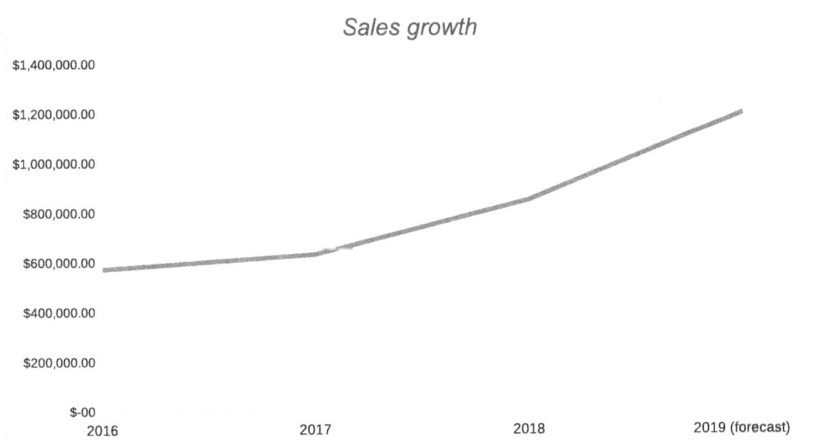

Figure 12.2 Epikus sales growth.

of managerial approaches to satisfy customers and event guests, dealing with service failures, etc.). The CIT is deemed reliable to obtain a comprehensive description of a specific content domain (Woolsey, 1986) and fits the objective of this study. Moreover, the CIT is a suitable research approach to gauge the effectiveness of certain managerial practices and to document professional roles that have an impact in service quality (Byrne, 2001).

During the semistructured interviews, interviewers focused on incidents that have contributed to the innovative capacities of the company, led to improvements in its abilities to provide memorable, holistic food experiences (Lashley, Morrison, & Randall, 2004; Lupton, 1994), and—due to the specific opportunities afforded by the personalized nature of boutique catering businesses—their implications for customer participation (Mustak et al., 2016) as well as value cocreation efforts (Helkkula, Kelleher, & Pihlström, 2012; Helkkula, 2011).

12.3 Findings and discussion

By the time interviews were concluded, an overall picture of the company's boutique concept and food service experience excellence emerged. Four major themes were identified in the empirical data: First interviewees communicated clear notions of the boundary between service and servility and articulated how they negotiate this boundary in their day-to-day encounters with customers. Second a theme that stood out as particularly relevant to the successful implementation of customer participation management was the interviewees' drive to contextualize new information and apply it to the service experience. Third with the service boundary-spanning roles aspects of the service experience in mind, social and relational aspects can greatly influence the customer's perception of the service experience. Facilitating social interaction is one of the staff's goals at the various stages of the service encounter. Finally the company was afforded specific innovative advantages due to its adjacent operations in fine dining and culinary education. These facilities provide competitive advantages as they foster the innovative capacities of staff by encouraging experimentation with culinary products and philosophies, as well as with methods of preparation and presentation. These capacities were further promoted by a positive and supportive company culture.

12.3.1 Communicating points of differentiation: service versus servility

The interviewees mentioned that first encounters with clients occasionally require educating them about points of differentiation from other catering businesses. These encounters can provide opportunities to shape customers' expectations about the service process and its outcomes (Mustak et al., 2016). To achieve these goals, patience and diplomacy are often required. One of the senior managers expressed his intuitions about shaping customer expectations in the following way: "I want to meet the person who coined the phrase 'the customer is always right' because I think the client does not always have good reasons to be right. In a polite way, you have to somehow make the customer understand that he or she is not acting in their best interest by telling us what to do. You have to respect their emotions and be diplomatic with difficult clients."

Some tensions can arise between the customer and the service provider due to the customer's inexperience with the differentiating aspects of boutique concepts and how their service delivery process incorporates customer participation. Communicating the differentiating aspects and processes that characterize innovative luxury service positioning are important steps to establish successful relationships with customers (Jin, Line, & Merkebu, 2016). The executive chef's following comment articulates how the company communicates these points of differentiation: "Beef filet is the basic bread and butter of all catering. And I decided that I would not serve it at any event and I would instead convince them [the customers] to consider more interesting choices. This is important to distinguish ourselves because I am not just selling the idea to the client at that moment, but also to their 200–300 guests at the event. So I had to make a very conscious decision and be controversial."

Overall, the interviews indicated that the benefits of customer participation borne of assuming a differentiating position in the catering service industry by far outweigh the risks; for instance, losing out on potential customers who simply shop for price and entirely discount experiential aspects.

12.3.2 Contextualizing the customer's experiences

As Pullman and Gross (2004) state, the proper understanding of context is of critical importance in the management and design of service experiences. Emerging service paradigms have received considerable interdisciplinary attention from marketing and hospitality researchers. The Service-Dominant Logic of Marketing (Vargo & Lusch, 2004, 2008, 2017) for instance suggests that value is always cocreated. Consequently service providers need to consider the beneficiary's history, environment, and current circumstances—in short, the beneficiary's context.

The executive chef of Epikus shared with us that he wishes to empower clients so that they can open up about their history and circumstances and more meaningfully participate in the creation of the service experience. The following quote articulates how Epikus' staff elicits customers' input and contextualizes it to translate it into an experiential service:

> *I would ask them some general questions. Especially the couple who would be getting married, trying to ignore the parents and the mother in law... mothers in general. And... I would ask the couples where have you traveled? Where did you study? What do you like to eat? Where do you have family? Some had family from Germany or Spain... or other countries and I try to incorporate culinary elements of those countries into the menu. [...] So, I would design a menu that reflects who the couple are, where they have been, what they have lived through, and of course what they like to eat. [...] You can extrapolate from their answers.*

During these initial meetings, customers often express specific ideas about the food they wish to be served at the event, but are often unaware of emotional needs, such as self-expression and validation. In the aforementioned situation, the chef gains knowledge of the customers' context, which can be translated into a culinary experience. The culinary cues are bound to be interpreted by the guests during the event, allowing them to better relate to the couple's life journey. Thus this arrangement incorporates an element of self-expression for the soon-to-weds into the event. The successful management of such intangible elements strongly influences customers' emotions and can positively affect the outcome of the service experience (Pullman & Gross, 2004).

12.3.3 Facilitating social interaction

The notions of innovation and technology are strongly intertwined, but of course innovation is entirely possible without technology. In fact, when we asked Epikus' PR manager about technology, it was clear from the answer that the firm was trying to reduce its role, at least in instances where it was mediating customer interactions: "Technology is cold, it is impersonal! We want to get to know our client personally, because they become a part of the event from the very beginning. If they come to our location we receive them very warmly and we get them involved, so the client feels as a part of the whole event planning."

Harking back to a point which the executive chef made about differentiating the company from the competition's offerings, he added:

> *Clients would always call in and ask for a menu... ask for a quote. But we found that that was not working for us because many menus look similar and the only thing the client would end up doing is compare providers according to their prices. So we decided not to work over the phone or internet. We decided to work in person with our clients when they called us. Please come meet us at the restaurant. And we would sit down, have a chat. That way we were able to communicate what exactly we represent in the industry.*

While these accounts show that the service staff puts the direct relationship with the customer front and center, the staff also seeks to facilitate social interaction among and with guests during events. The executive chef shared observations with us that he had made and how he contextualized these observations to improve Epikus' service experience by reducing formality among guests:

> *What I noticed when I was invited to weddings is that the biggest and best weddings... supposedly... they would have all this excessive decoration. So... they would have these huge flower arrangements in the center of the table. And... it's like... you're trying to have a conversation trying to look at people around that thing [leans left and right demonstrating the movement to establish eye contact]... it's just very awkward. So... we have these very wealthy clients and they want to see or copy what they've seen before. And... I tell them why don't you just throw that huge flower arrangement in the garbage? Instead, I suggest we put tapas in the center of the table and they'd share food... and... the food becomes a*

> *talking point... and in the end it reduces the formality... and people relax. And they remember this sort of thing....*

While perhaps more cumbersome, low-tech communication channels provide a greater bandwidth to emotionally connect with customers and to communicate the image of the company in a richer way than marketing collateral possibly could. The importance of relationships during the entire service experience is further emphasized by the efforts to reduce formality to facilitate positive social interactions among guests.

12.3.4 Entrepreneurial marketing and innovative capacities

Entrepreneurial marketing approaches have been associated with the successful marketing practices of boutique concepts (Kurgun et al., 2011; Morris, Schindehutte, & LaForge, 2002). Among other things, entrepreneurial marketing firms embrace calculated risk taking and experimentation to foster their innovative capacities. Our interviewees pointed out synergies between the catering business and the adjacently operating culinary school. The PR manager stated:

> *Our fine dining and catering services have a strong academic component, which helps us in many ways to be innovative. We have to be at the forefront of absolutely everything. We have students coming from all over the world who are involved in the latest trends...and they are always eager to learn and seeking out new innovative methods of food preparation or presentation... or cooking philosophies. Picture for instance a classic French preparation like Pommes de Terre à la Lyonnaise, but with vibrant and colorful Andean tubers instead of potatoes. Our teachers and students want to experiment, adapt dishes and always try to incorporate new ingredients. That is why our menu is never fixed and we can offer our students and chefs the freedom to author their own signature menus.*

While the aforementioned statement articulates how the academic component of the business fosters innovation by encouraging experimentation, it also introduces managerially challenging elements as one of the senior managers points out:

> *It's a bit of a cliché, but making mistakes is often the best learning experience for our students... [chuckles] ... although we try to avoid those when clients are around. [...] Since our students and teaching chefs add a certain dynamism to our operations it is not really possible for us to offer the ahh... set in-stone, prenegotiated sort of package-deal catering menus that many of our competitors use. We like to negotiate this day by day... and we have to adapt to what we find on markets and what we want to offer our clients. So .. sometimes that can be a disadvantage in terms of profit margin, but it also usually gives us an edge in terms of differentiation.*

Ultimately the positive and innovation-fostering values of entrepreneurial marketing approaches need to be weighed against their potential downsides.

12.4 Practical implications

Successful boutique caterers acknowledge the complexity of the process and the variability of tangible and experiential resources that they integrate into a unique, high-quality, and holistic experience.

When customers seek out caterers, they often carry in their minds a mental representation of the event that they desire. It is crucial to train staff to absorb and interpret these representations and to translate them into successful service experiences. While customers' ideas about the food and its presentation are made explicit, the customers are often unaware of the emotional needs that they seek to fulfill at the same time (e.g., self-expression and validation). Frontline employees require sensitivity to these needs and a diplomatic communication style to situate customers' wishes in the appropriate culinary context and suggest appropriate variations.

Potential conflicts of interest need to be carefully managed during this initial get-to-know phase. For instance, if a customer insists on a venue that could potentially compromise the company's brand perception and reputation among guests, standing one's ground and declining a customer's inquiry is perhaps the better option. In many instances it is practical for boutique caterers to form mutually beneficial strategic partnerships with supporting companies (e.g., providers of location, décor, entertainment, and lighting). It is important to align all the elements that comprise the event with the type of experience the caterer and the planner intend to create. Therefore strategic partners that supply supporting services such as location, décor, entertainment, and lightning, need to comply with the standards, the style and the attention to detail required to stage that event. Otherwise an unaligned provider can harm the perception of other service providers.

In contrast to the controllable environment of a restaurant, catering companies have to take into account many unknown variables and plan dynamically. Detailed preproduction and location scouting are key elements of the initial stages. It is self-evident that customers expect a certain quality of the consumables, but orchestration of their timely delivery and quality of presentation utensils (e.g., glassware, silverware, napkins, and tablecloths) further leverages customer satisfaction.

Throughout the planning and execution of an event, material aspects such as the visual presentation of consumables are interpreted by the customer and guests. Additionally nonmaterial aspects, such as the self-presentation, demeanor, and communication skills of wait staff and bartenders, send signals about the quality and uniqueness of the event and should not be neglected at any stage.

Each customer inquiry is a new challenge, and—while it is unavoidable to repeat culinary themes—marketing communications should be tailored to avoid the portrayal of recurring menus or dishes. Customers' "crazy ideas" should be seized as opportunities by meeting them with openness and translating these ideas into successful events as they can generate extraordinarily positive word of mouth and social media engagement.

Interviews have been edited for brevity and clarity.

Appendix A Food presentation

Appendix B Service experience

References

Aggett, M. (2007). What has influenced growth in the UK's boutique hotel sector? *International Journal of Contemporary Hospitality Management*, *19*(2), 169–177.

Bott, G., & Tourish, D. (2016). The critical incident technique reappraised. *Qualitative Research in Organizations and Management: An International Journal*, *11*(4), 276–300.

Byrne, M. (2001). Critical incident technique as a qualitative research method. *AORN Journal*, *74*(4), 536–539.

Chathoth, P. K., Ungson, G. R., Harrington, R. J., & Chan, E. S. W. (2016). Co-creation and higher order customer engagement in hospitality and tourism services: A critical review. *International Journal of Contemporary Hospitality Management*, *28*(2), 222–245.

Cope, J., & Watts, G. (2000). Learning by doing – An exploration of experience, critical incidents and reflection in entrepreneurial learning. *International Journal of Entrepreneurial Behavior & Research*, *6*(3), 104–124.

Flanagan, J. C. (1954). The critical incident technique. *Psychological Bulletin*, *51*(4), 327.

Gummesson, E. (2008). Extending the service-dominant logic: From customer centricity to balanced centricity. *Journal of the Academy of Marketing Science*, *36*(1), 15–17.

Helkkula, A. (2011). Characterising the concept of service experience. *Journal of Service Management*, *22*(3), 367–389.

Helkkula, A., Kelleher, C., & Pihlström, M. (2012). Characterizing value as an experience. *Journal of Service Research*, *15*(1), 59–75.

Jin, N. (Paul), Line, N. D., & Merkebu, J. (2016). Examining the impact of restaurant innovativeness on relationship quality in luxury restaurants. *International Journal of Hospitality & Tourism Administration*, *17*(4), 449–471.

Jones, D. L., Day, J., & Quadri-Felitti, D. (2013). Emerging definitions of boutique and lifestyle hotels: A Delphi study. *Journal of Travel & Tourism Marketing*, *30*(7), 715–731.

Kurgun, H., Bagiran, D., Ozeren, E., & Maral, B. (2011). Entrepreneurial marketing—The interface between marketing and entrepreneurship: A qualitative research on boutique hotels. *European Journal of Social Sciences*, *26*(3), 340–357.

Lashley, C., Morrison, A., & Randall, S. (2004). My most memorable meal ever! Hospitality as an emotional experience. *Culinary Taste*, 165–184.

Lupton, D. (1994). Food, memory and meaning: The symbolic and social nature of food events. *The Sociological Review*, *42*(4), 664–685.

Lusch, R. F., & Vargo, S. L. (2006). Service-dominant logic: Reactions, reflections and refinements. *Marketing Theory*, *6*(3), 281–288.

Mcintosh, A. J., & Siggs, A. (2005). An exploration of the experiential nature of boutique accommodation. *Journal of Travel Research*, *44*(1), 74–81.

Morris, M. H., Schindehutte, M., & LaForge, R. W. (2002). Entrepreneurial marketing: a construct for integrating emerging entrepreneurship and marketing perspectives. *Journal of Marketing Theory and Practice*, *10*(4), 1–19.

Mustak, M., Jaakkola, E., Halinen, A., & Kaartemo, V. (2016). Customer participation management. *Journal of Service Management*, *27*(3), 250–275.

Prahalad, C. K., & Ramaswamy, V. (2004). Co-creation experiences: The next practice in value creation. *Journal of Interactive Marketing*, *18*(3), 5–14.

Pullman, M. E., & Gross, M. A. (2004). Ability of experience design elements to elicit emotions and loyalty behaviors. *Decision Sciences*, *35*(3), 551–578.

Schembri, S. (2006). Rationalizing service logic, or understanding services as experience? *Marketing Theory*, *6*(3), 381–392.

Shaw, G., Bailey, A., & Williams, A. (2011). Aspects of service-dominant logic and its implications for tourism management: Examples from the hotel industry. *Tourism Management*, *32*(2), 207–214.

Thompson, J. D. (1962). Organizations and output transactions. *American Journal of Sociology*, *68*(3), 309–324.

Vargo, S. L., Maglio, P. P., & Akaka, M. A. (2008). On value and value co-creation: A service systems and service logic perspective. *European Management Journal*, *26*(3), 145–152.

Vargo, S. L., & Lusch, R. F. (2004). Evolving to a new dominant logic for marketing. *Journal of Marketing*, *68*(1), 1–17.

Vargo, S. L., & Lusch, R. F. (2008). Service-dominant logic: Continuing the evolution. *Journal of the Academy of Marketing Science*, *36*(1), 1–10.

Vargo, S. L., & Lusch, R. F. (2017). Service-dominant logic 2025. *International Journal of Research in Marketing*, *34*(1), 46–67.

Victorino, L., Verma, R., Plaschka, G., & Dev, C. (2005). Service innovation and customer choices in the hospitality industry. *Managing Service Quality: An International Journal*, *15*(6), 555–576.

Woolsey, L. K. (1986). The critical incident technique: An innovative qualitative method of research. *Canadian Journal of Counselling and Psychotherapy/Revue canadienne de counseling et de psychothérapie*, *20*(4).

Curated nostalgia—constructing and consuming the city through a culinary adventure

13

Tony Stovall[1], *Mine Üçok-Hughes*[2] *and Rossen Ventzislavov*[3]
[1]Kelley School of Business, Indiana University, Indianapolis, IN, United States, [2]College of Business and Economics, California State University, Los Angeles, Los Angeles, CA, United States, [3]College of Liberal Arts, Woodbury University, Burbank, CA, United States

13.1 Introduction

In this chapter, we examine how the unique relationship between food and the city helps shape what we consider a negotiation between often conflicting narratives. There is one constant that remains true for every city—they all undergo physical changes over time. But there is a dimension of these changes that is much less immediate and much harder to articulate—the city's ever-evolving symbolic identity. Interestingly the symbolic identity of the city sometimes communicates a counterpoint to its physical presence, as is the case with Los Angeles, where only 3% of the population serves the city's symbolic purpose—the business of entertainment (Kerr, n.d.).

It is within this conflicted context that the thematic dining series, *Los Angeles Eats Itself*—the subject of this case study—becomes emblematic of the city's collective narrative identity and food consumption practices. *Los Angeles Eats Itself* exploits this narrative ambiguity to create "a dinner series where cuisine and LA noire merge like freeway onramps in a savory digestible history" (Los Angeles Eats Itself, n.d.). Ultimately we suggest that our literal and figurative taste of and for Los Angeles is in part a function of these adventures and the stories that drive them. We explore culinary adventure as a driver for the complex set of relationships between food, what we term "curated nostalgia," and identity—the personal identity of the particular urban dweller and the symbolic identity of the city taken as a dynamic totality. We see this relationship between the urban dweller and the city as a negotiation between narratives.

This chapter is structured as follows: first, the chapter discusses the authors' conceptualization of the city as both a story and an object to be consumed. Next, it incorporates the ways in which food adds to our collective identification with the city and the role of nostalgia in creating this bond. The chapter continues with a detailed description of The Fleiss Feast—an iteration of the dining series—followed by a brief conclusion with an analysis of the way stories and food shape our nostalgia for the city.

Case Studies on Food Experiences in Marketing, Retail, and Events. DOI: https://doi.org/10.1016/B978-0-12-817792-1.00013-7

13.2 The city as both a story and an object to be consumed

13.2.1 Consuming the city as a story

In his discussion of the "pleasure of 'seeing the whole', of looking down on, totalizing the most immoderate of human texts," Certeau (1984) admits two fundamental aspects of our relationship to the city—our tendency to narrate the urban environment as a story and our fantasy about apprehending "the whole picture" all at once (Certeau, 1984, p. 92). According to Certeau, the urban narrative "makes the complexity of the city readable and immobilizes its opaque mobility in a transparent text" (Certeau, 1984, p. 92). The symbolic is the aspect that we capture and articulate through storytelling (Benjamin, 1999; Hughes, 1975). Our narrative engagement with the city is so prevalent that, according to Certeau, even our physical movement through the urban environment charts a "space of enunciation" (Certeau, 1984, p. 98). By this we mean that the unique trajectories residents and visitors traverse—commutes, diversions, explorations—are not derived from any city's identity, in this case Los Angeles, but are constitutive of it.

13.2.2 Consuming the city as an object

The geographical peculiarities and spatial realities of a city do as much to explain the consumption behaviors, especially concerning food, of those who reside within its boundaries as it does to explain the behavior of those who reside without (Harvey, 1989). Nevertheless, communities become embedded within one another, making it difficult to disentangle them. It is this embeddedness that facilitates the creation of collective identities.

Those who have a favorable attitude toward their city of residence will more closely identify with it. Freire (2005) argued that inhabitants of a city are but mere consumers of it, consumers of the goods, services, and experiences in the city, and this act of consumption is crucial to one's sense of belongingness in that community. This belongingness cognitively comprises the "memories, ideas, feelings, attitudes, values, preferences, meanings, and conceptions of behavior and experiences which relate to the variety and complexity of physical settings that define the day-to-day existence of every human being" (Proshansky, 1978). The experiential aspects of consumption include phenomena such as playful leisure activities, sensory pleasures (restaurants, food festivals, etc.), daydreams, aesthetic enjoyment, and emotional responses (nostalgia, etc.) (Holbrook & Hirschman, 1982), all of which a city can potentially offer its residents.

13.3 Food as a system of communication

Food is one of the many ways in which people consume a city and establish a relationship with it. In *An Edible History of Humanity*, Tom Standage states that "throughout history, food has done more than simply provide sustenance. It has acted as a catalyst of social transformation societal organization, geopolitical competition, industrial development, military conflict, and economic expansion" (Standage, 2009, p. ix). Over time food has become its own system of communication (Barthes, 1997 [1961]; Douglas, 1991) that in many ways has come to reflect the interactions among members of particular social and cultural groups (Ucok Hughes et al., 2017).

Lévi-Strauss (1997) proposed that cooking is a cultural transformation of nature and can be used to analyze social behavior across a range of societies because it is an intimation of universally shared mentalistic structures. In "Deciphering a Meal" (1991) Douglas claims that "if food is treated as a code, the messages it encodes will be found in the pattern of social relations being expressed" (p. 249). In his essay "Towards a Psychosociology of Contemporary Food Consumption" Barthes (1997 [1961]), argues that food is a system of communication, signifies cultural meanings, and eventually evokes nostalgic associations in those who consume it (Wood, 1995).

13.4 Nostalgia for the city, curated nostalgia, and culinary adventure

Kessous and Roux (2008) suggested that nostalgia can play an important role in creating and maintaining identity through various life cycles and across time and space. And despite its status as a secondary emotion, lacking the strength of primary emotions like hate and grief, nostalgia, like food traditions, is also culturally derived and connects one to valued past experiences and specific locales and associated participants. Nostalgia has a unique ability to assist in the cocreation of experiences, as well as allowing us to vicariously experience the past through the memories of others and by utilizing "imagined fabrication" of the importance of objects, and in the case of *Los Angeles Eats Itself*, food, to the specific time period (Jafari & Taheri, 2014). It is both personal and social and wildly discursive in nature (Higson, 2014).

Oxforddictionaries.com defines nostalgia as, "a sentimental longing or wistful affection for a period in the past." Given the etymology of nostalgia from the Greek words—*nostros* (homecoming) and *algos* (ache), the concept has always suggested geographical roots (Vignolles & Pichon, 2014). Memories are contextually situated and culturally predicated (Olick, 1999) and are formed in relation to and in communication with others, and can be instrumental in bridging otherness through a common understanding of the past. Furthermore they create a cultural heritage which helps to situate and individuate the respective culture within the larger constellation of cultures (Assmann & Czaplicka, 1995).

Los Angeles Eats Itself describes itself as "a dinner series where cuisine and LA noir merge like freeway onramps into a savory digestible history. As food rapidly becomes an aesthetic experience of choice, cities are increasingly seen through the lens of their edible itineraries" (Los Angeles Eats Itself, n.d.).

13.5 Case study context, background, and methodology

In an in-depth interview with Jason Keller, one of the creators of the dinner series, conducted shortly after two authors attended *The Fleiss Feast*, he states that "the idea was not to glorify or celebrate these gory and salacious events but they were all parts of a shared narrative that brought the culture, people living in LA together. They were all unifying experiences" (Keller, 2016).

All three authors participated in two different culinary events as participant observers (*The Black Dahlia Dinner* [Ventzislavov attended, April 15, 2015] and *The Fleiss Feast* (Stovall and Üçok Hughes attended, March 6, 2016)). During the *Los Angeles Eats Itself* dinner events authors took photos of the food, participants, décor and different aspects of the events in general. In a mini "welcome" talk to open *The Fleiss Feast*, the event creators provided the contextual and symbolic meanings of each menu item and the décor of the event venue. To gauge whether others who attended the events perceived the events as the curators intended, the case study authors searched online for press mentions and relevant articles by using the key phrase "Los Angeles Eats Itself." All events in the series had the attention of the local press yielding a number of articles.

This case study specifically focuses on *The Fleiss Feast*, one in a series of thematic dining events focused on "many of LA's most notorious moments [that] border on the tragic, catastrophic and macabre" (Los Angeles Eats Itself, n.d.). *The Fleiss Feast* was held in the evening of March 6, 2016, at a rented indoor studio space in Los Angeles. The dinner event and the menu were deliberately curated to evoke nostalgia of the salacious disquietude surrounding the "Hollywood Madame," Heidi Fleiss, convicted of and sentenced to prison for operating a prostitution ring in the 1990s.

Upon entering the venue, hit songs reminiscent of the 1990s provided an ever-present musical backdrop. The guests were greeted by a host, in a makeshift reception area, and asked to sign in a black book evoking Heidi Fleiss' infamous "Black Book" that contained the names of her clients. After "registering," they were each given a gold-plated plastic fork attached to a plastic hotel room key imprinted with their seat number. During the "cocktail hour," guests were served beverages fashioned after Zima, a brand of lightly alcoholic beverage popular in the 1990s, coupled with an appetizer named Gilded Girl—gold-encrusted potato skins stuffed with lobster mash. This clash between a relatively inexpensive vegetable and its glamorous presentation was meant to evoke the duplicity of Fleiss' brand—a veneer of glamour barely concealing a world of depravity.

The setting of this dinner was a "dining hall" decorated to resemble the poolside at the Beverly Hilton Hotel where Fleiss held many of her business meetings. An exact replica of the wallpaper from the Beverly Hilton decorated a wall guests used as a backdrop for the photos they took, some of which ended up on Instagram (see #fleissfeast, #losangeleseatsitself). The dinner tables, each set under umbrellas and artificial palm trees were placed around the projected image of a pool created by artist Christopher Reynolds. "With a virtual shimmer, guests lounged around the holographic water while listening to a dance mix that only a 90's child could appreciate" (The Fleiss Feast, n.d.). The "pool" was created by a light projected onto a framed-off space with no depth in the middle of the studio floor.

A round mirror served as a placemat accompanied by a thin, long metal tube, whose function became apparent later on. Functionally it was meant to facilitate the consumption of the soup course; figuratively, it connoted the flagrant consumption of cocaine among the glitterati that comprised her clientele. On each seat was an envelope containing the menu and a parrot feather, alluding to Fleiss' later-life interest in parrots, and a pink hypercolor napkin. The pink was a color typical for the 1990s and, also, for beach bodies. When exposed to body heat it turned white in patterns reminiscent of the acid-wash denims that were highly popular in the 1990s. Also the idea was that the light pink napkin laid across one's crotch (erogenous zone) had the potential to turn white if the person got too hot in that area. Moreover, the napkins were washed in Fleiss' very own *Dirty Laundry* laundromat in Las Vegas.

Other references to Fleiss' life, business, and the culture of the era included a collection of *eau de colognes* popular in the 1990s that were placed in the bathrooms where *Alien 51*, a film Fleiss starred in was shown on a monitor while the biopic *Call Me: The Rise and Fall of Heidi Fleiss* ran, with the sound off, in the reception area.

The dinner began with the cocreators presenting the carefully curated menu. Each dish was developed to reference specific events and practices associated with Heidi Fleiss, as well as evoke a particular sense of time and place. In preparing the menu the artists and the chefs wanted each dish to be its own "talking point" (Keller, interview June 2, 2016).

At the time of our writing, four dinners out of a planned ten had been held between March 2015 and April 2019. See Appendix A for a full description of the *Los Angeles Eats Itself* concept. Appendix B provides descriptions of the dinners held up to the point of this case study.

13.6 Fleiss Feast menu

The Fleiss Feast menu items, their descriptions and symbolic meanings intended by the creators of the event are presented below. Photographs of the dishes describes can be seen at http://christopherreynoldsstudio.com/the-fleiss-feast.

- *Paparazzi popcorn*: The paparazzi phenomenon became big in the 1990s. The idea of relentless media coverage emerged during this time. Heidi was partly complicit in the emergence of that culture.
- *The Hollywood Madame*: Oyster, sparkling wine, passion fruit mignonette. Oysters are believed to be an aphrodisiac while the name "passion fruit" readily lends itself to lurid interpretations.
- *The Sting*: Smoky ham bone broth poured over powdered buttermilk "cocaine" that turns into a clear soup. The idea was that one would sip the soup through the long thin metal straw bent over the bowl and the mirror underneath it - a reference to the drug culture of the 1990s.

- *Midnight Caesar*: Squid ink croutons, black garlic dressing, white anchovies. The black color is a reference to Fleiss' Black Book and the dark side of her business enterprise.
- *Parrot caviar*: A small tin can filled with beluga lentils, foie mousseline, and a quail egg all topped with fennel sprigs. The lentils and the quail eggs were chosen for their resemblance to caviar parrot eggs respectively, the latter an oblique reference to Fleiss' fascination with parrots. While fascinating from a culinary perspective, these choices were meant to remind diners of a harsh Hollywood reality—the fact that there is often very little behind the facades people and places present to us.
- *Surf & turf*: A cube of beef cheek topped with a king prawn and red pepper sauce. Surf and Turf was a very popular dish in the 1990s which combines seafood and red meat. There is also a wordplay winking at the "turf wars" between Madam Alex (Fleiss' ex-boss) and Heidi Fleiss.
- *Cherries & chocolate*: Chocolate meringue molded into vagina shapes and topped with an alcohol-soaked cherry accompanied with house-made push pops.

13.7 Discussion

The idea for this dinner series came out of a *New York Times* article (Deresiewicz, 2012) that argued that food has begun to replace art as a city-defining tool. "Foodism has taken on the sociological characteristics of what used to be known — in the days of the rising postwar middle class, when Mortimer Adler was peddling the Great Books and Leonard Bernstein was on television — as culture." Interestingly Los Angeles, colloquially referred to as "The City of Angels," "Tinseltown" and "La La Land," despite being known as a progenitor of narratives has no singular narrative itself. "*Los Angeles Eats Itself* is an effort to present a range of narratives in a unique and tangible way" (Los Angeles Eats Itself, n.d.).

The event's literal and symbolic reconstruction of a historical moment occasioned an amalgamation of narratives. It is out of this amalgam that the participants created a new "negotiated identity" (Ucok Hughes et al., 2017) based on their individual experiences in relation to the 1990s in general, and the Heidi Fleiss phenomenon in particular. For instance, while particular hit songs from the 1990s (playing in the background throughout the event) evoked certain feelings for the participants of the event, the symbolic significance of the dish "surf & turf" was lost to some who were either not old enough to remember or were not living in the United States to experience the prominence of this dish on the restaurant scene. Similarly for some, the tabloid headlines of the day Fleiss was arrested brought back vivid memories, while not resonating with others. Instead of alienating anyone, those "blind spots" created a sense of belonging—like blank spaces in a collective script all participants were tasked to collectively reenact.

13.8 Managerial implications

Each participant likely had wildly different associations with and/or memories of Fleiss and her public unraveling. This grasp of nostalgia, however, came into effect as a result of the partial evocations of a historical past that we all turned out to have miraculously shared. Nostalgia has the unique ability to assist storytelling, in the cocreation of experiences, as well as allowing us to vicariously experience the past through the memories of others and by utilizing "imagined fabrication" of the importance of objects, and in the case of *Los Angeles Eats Itself*, food, to the specific time period (Jafari & Taheri, 2014). Our divergent visions conspired to create a situational assemblage that articulated 1990s American life and the role Los Angeles played in it.

Storytelling plays an important role in the formation of both personal and collective identities. Scholars like Baudelaire, Benjamin, and Certeau transposed the narrative mode of identification onto the urban fabric. Through this case study the authors argue that narrativity is as helpful in understanding our consumption of the city as it is in articulating our phenomenological and semiotic engagements with it. *Los Angeles Eats Itself* as a culinary adventure series provides a distinctive lens through which to examine the role of food in the construction of these narrative identities. Culinary adventure, curated or otherwise, is a phenomenon specific enough to help illustrate the conceptual implications of "curated nostalgia." *Los Angeles Eats Itself* in particular is a pertinent example of the productive tension between dueling narratives. We explore this tension through the dinner series that recreates and reconsumes ignominious Los Angeles stories. The fact that *Los Angeles Eats Itself* is a curated, scripted event makes it easier to illustrate this balance. Each city, with its own unique history, can follow the road map created by *Los Angeles Eats Itself* to create its own dinner series. This dinner series does not simply sensationalize dark episodes of the city's past—it uses the thrill of prurient events to help us reexamine our relationship with the city's identity, and the myriad ways we consume it.

References

Assmann, J., & Czaplicka, J. (1995). Collective memory and cultural identity. *New German Critique*, *65*, 125–133.

Barthes, R. (1997). Toward a psychosociology of contemporary food consumption. In C. Counihan, & P. Van Esterik (Eds.), *Food and culture: A reader*. New York, London: Routledge.

Benjamin, W. (1999). *The arcades project*. Cambridge: Harvard University Press, Howard Eiland & Kevin McLaughlin, Trans.).

Certeau, D. M. (1984). *The practice of everyday life*. Berkeley: University of California Press, Steven Rendall, Trans.).

Deresiewicz, W. (2012, October 26). A matter of taste? *New York Times*. Available from <http://www.nytimes.com/2012/10/28/opinion/sunday/how-food-replaced-art-as-high-culture.html?_r = 1>.

Douglas, M. (1991). *Deciphering a meal. Implicit meanings: Essays in anthropology*. London, New York: Routledge.

Freire, J. R. (2005). Geo-branding, are we talking nonsense? A theoretical reflection on brands applied to places. *Place Branding*, *1*(4), 347–362.

Harvey, D. (1989). *The urban experience*. Baltimore, MD: Johns Hopkins University Press.

Higson, A. (2014). Nostalgia is not what it used to be: Heritage films, nostalgia websites and contemporary consumers. *Consumption, Markets & Culture*, *17*(2), 120–142.

Holbrook, M. B., & Hirschman, E. C. (1982). The experiential aspects of consumption: Consumer fantasies, feelings, and fun. *Journal of Consumer Research*, 132–140.

Hughes, H. S. (1975). *History as art and science*. Chicago, IL: University of Chicago Press.

Jafari, A., & Taheri, B. (2014). Nostalgia, reflexivity, and the narratives of self: Reflections on Devine's "Removing the rough edges?". *Consumption, Markets & Culture*, *17*(2), 215–230.

Keller, J. Personal interview June 2, 2016.

Kerr, J. (n.d.) *Concentration of entertainment-related employment by metropolitan area. Occupational area statistics*. United States Department of Labor, Bureau of Labor Statistics. Available from <https://www.bls.gov/oes/entertainment.pdf> Accessed 02.20.17.

Kessous, A., & Roux, E. (2008). A semiotic analysis of nostalgia as a connection to the past. *Qualitative Market Research: An International Journal*, *11*(2), 192–212.

Lévi-Strauss, C. (1997). The culinary triangle. In C. Counihan, & P. Van Esterik (Eds.), *Food and culture: A reader*. New York, London: Routledge.

Los Angeles Eats Itself. (n.d.). <http://losangeleseatsitself.com/>.

Olick, J. K. (1999). Collective memory: The two cultures. *Sociological Theory*, *17*(3), 333–348.

Proshansky, H. M. (1978). The city and self-identity. *Environment and Behavior*, *10*(2), 147–169.

Standage, T. (2009). *An edible history of humanity*. Bloomsbury Publishing USA.

The Fleiss Feast. (n.d.). *Los Angeles eats itself*. Retrieved from <losangeleseatsitself.com/event/madam-heidi-fleiss-feast> Accessed 02.20.17.

Ucok Hughes, M., Stovall, T., & Ventzislavov, R. (2017). Constructing a narrative identity of Los Angeles through a gastronoir adventure. In A. Gneezy, V. Griskevicius, & P. Williams (Eds.), *NA – Advances in consumer research* (Vol. 45, pp. 486–488). Duluth, MN: Association for Consumer Research.

Vignolles, A., & Pichon, P. E. (2014). A taste of nostalgia: Links between nostalgia and food consumption. *Qualitative Market Research: An International Journal*, *17*(3), 225–238.

Wood, R. (1995). *The sociology of the meal*. Edinburgh: Edinburgh University Press.

14 Experiences gained through absorption of and immersion in destinations' food events—case evidence from Finnish travellers

Peter Bjork[1] and Hannele Kauppinen-Räisänen[2]
[1]Department of Marketing, CERS, Hanken School of Economics, Vaasa, Finland,
[2]School of Marketing and Communication, University of Vaasa, Vaasa, Finland

14.1 Introduction

People's daily activities greatly change during vacations and holidays away from home. Nevertheless, as one of the basic human needs, eating remains a vital part of travellers' daily routines. Thus travellers' experience changes according to the types of foods and drinks consumed and the place and time of consumption.

In their recent study, Björk and Kauppinen-Räisänen (2019) hold that travellers typically consume food at various events. These situations vary according to the environment, whether these are organized or nonorganized, the participants at the event and whether the participants are travellers or residents. Food experiences emerge in organized environments, such as hotel areas, bars, and farms, which are particularly staged as tourism attractions and intended for travellers (Björk & Kauppinen-Räisänen, 2019). On the contrary, other organized places, such as retail stores, market squares, and food markets, are established to serve primarily the needs of local inhabitants. Experiences also evolve in nonorganized environments, which are mostly used by travellers (e.g., accessible beaches and streets) and locals (e.g., at residents' homes and remote and inaccessible [for tourists] beaches, streets and roadsides).

The experiences are driven by contemporary travellers' interest in what food various destinations can provide, such as local ingredients, flavors and food products and traditional but exotic dishes (e.g., Long, 2010). Previous studies that focus on food experiences reveal that travellers' personal interests and motivations in regard to food affect their experiences with food during their holidays abroad (e.g., Björk & Kauppinen-Räisänen, 2016a, 2016b). Research also shows that travellers greatly desire experiencing changes to one's everyday life, which destinations' food responds to (Björk & Kauppinen-Räisänen, 2016a, 2016b; Richards, 2002). Hence, participation in destinations' food events allows food experiences to evolve (Long, 2010). Some of these experiences are common, whereas some are extraordinary to a certain extent (Björk & Kauppinen-Räisänen, 2016b). Regardless, a destination's

Case Studies on Food Experiences in Marketing, Retail, and Events. DOI: https://doi.org/10.1016/B978-0-12-817792-1.00014-9

culture and unique cuisine may trigger travellers' travel motivation and determine their destination choice (e.g., Björk & Kauppinen-Räisänen, 2019; Boniface, 2003; Hall & Sharples, 2003; Hsu, Tsai, & Wu, 2009; Kivela & Crotts, 2005). Although past research does not provide evidence of the difference between various food events' contribution to food experiences, the importance of a destination's foodscape and its various food events are well acknowledged in the literature. That is, past research shows how food experiences contribute to travel satisfaction, revisit intentions and a holistic travel experience and holistic travel experience (e.g., Björk & Kauppinen-Räisänen, 2017a, 2017b; Kim, Kim, Goh, & Antun, 2011; Neild, Kozak, & LeGrys, 2000).

This study focuses on a destination's foodscape, which comprises various food events. The study is based on two basic premises. First, the study is based on the idea that travellers may play various roles at destinations' food events. These events contribute to the traveller's experiences, which may vary according to whether the traveller participates actively or passively in an event. According to the idea by Pine and Gilmore (1999), travellers may take part in events by playing various roles. That is, the traveller may take a passive role and absorb information or take an active, participatory role and immerse him/herself in a food event (Carù & Cova, 2003; Pine & Gilmore, 1999). Yet, this issue has received little scholarly attention within the research on travellers' food experiences. Second, the study is based on the notion that destinations' events stimulate the basic senses of humans, which ideally contribute to good food experiences. Related to that past research has contributed with insights into the internal or intrinsic aspects of food, such as food's locality and traditionality. Many studies show how these aspects are essential features of a destination's food culture and cuisine as the contemporary traveller perceives that these food aspects contribute to good experiences (Björk & Kauppinen-Räisänen, 2016b; Hjalager & Richards, 2002; Okumus, Okumus, & McKercher, 2007; Quan & Wang, 2004). The recent concept of gastrophysics holds that food experiences are also composed of such aspects as the color of the crockery on which food is served, the weight of the cutlery and the sound and lighting at the event where food is consumed (Spence, 2018). Indeed, research within a given foodscape, dinescape, and platescape supports the importance of the external aspects of food and the idea that basic senses affect food experiences (Cummins & Macintyre, 2002; Mikkelsen, 2011; Ryu, 2005; Sobal & Wansink, 2007). Various food events engage travellers in different manners, and it can be assumed that the basic senses are stimulated particularly in food encounters which are immersive. Yet, the external or extrinsic aspects of food that most likely also contribute to travellers' food experiences have received less scholarly attention.

The previous discussion implies that human-related aspects and foods' external aspects need careful attention as they also influence travellers' food experiences. Hence, an interesting question is, are all the stimulated senses important for food experiences, including in cases where the traveller takes a passive role by observing a food event? To answer this question, the present research takes a case study approach and explores the various roles of travellers at destinations' food events and the importance of humans' basic senses in travellers' food experiences. The case evidence is gathered from among Finnish travellers. The study investigates the following:

1. How destinations' food events affect travel intention?
2. How a traveller's role relates to events that contribute to their food and eating experiences?
3. How a traveller's role relates to humans' basic senses that affect travellers' food experiences at various events?
4. How destinations' food events add to travel satisfaction?

14.1.1 Experiences lived through absorption and immersion

A destination's food culture and unique cuisines are made of local ingredients, flavors, and food products as well as traditional but exotic dishes. Such characteristics are acknowledged within the field of gastronomy, which stresses not only food culture and uniqueness but also highlights the role of traditions, food preparation, and cooking as well as rules and norms related to eating and drinking (Santich, 2004; Hegarty & O'Mahony, 2001). Sims (2009) specifies that destinations—whether entire countries or local areas—are identified by their gastronomy. Indeed, tourism destinations are identified by their gastronomy and attached to unique dishes; for example, Greece is identified by feta cheese and colorful salads, Japan is known for sushi, and Britain is attached to fish and chips.

The food culture and unique cuisine contribute to the appeal of a destination's foodscape and its various events. These events allow experiences to evolve, some of which are common, whereas others are extraordinary to a certain extent (Björk & Kauppinen-Räisänen, 2016b). As they vary from everyday experiences at home, a destination's foodscape and associated events are able to attract the attention of various types of travellers. Although food events such as farms and food markets may originally have started to serve the needs of local inhabitants, the fact is that the various events at destinations are increasingly acknowledged as essential tourist attractions (de Jong & Varley, 2017; Frisby & Getz, 1989; Mason & Paggiaro, 2012). These events have transformed from basic community service encounters to tourism attractions, which also provide means for destinations to develop (Getz, 1997).

The various events at destinations and travellers' participation in and connection to them provide various options, one being that not all travellers are involved and committed and wish to be active in an event (Björk & Kauppinen-Räisänen, 2017b).

Pine and Gilmore (1999) explain how consumers can take different positions on service encounters. They have suggested that participation can be categorized into active and passive participations. First a traveller may participate passively and thus does not affect the event. In such cases, the traveller is rather an observer at the event. For example, the traveller may watch farm and food demonstrations at a food market from a distance or just pass by (Quadri-Felitti & Fiore, 2012). Second the traveller may participate actively at an event by actively creating or cocreating an event, for example, by eating local dishes and tasting local ingredients and products (Kauppinen-Räisänen, Gummerus, & Lehtola, 2013; Organ, Koenig-Lewis, Palmer, & Probert, 2015). In these scenarios, the traveller personally affects the event compared with passive participation, where the traveller does not directly affect the event. Pine and Gilmore (1999) have also asserted that experiences can be described based on how the traveller connects to the event. These dimensions are defined as follows. Absorption means "occupying a person's attention by bringing the experience into the mind," and immersion is "becoming physically (or virtually) a part of the experience" (Pine & Gilmore, 1999, p. 31). Hence, the traveller can either absorb or immerse him/herself in the event. Travellers observing a food event absorb the event, whereas those living the event with all their basic senses are immersed in the event. Therefore absorption implies taking an interest in an event, whereas immersion means merging deeply within the event.

Pine and Gilmore (1999) defined four types of experiences based on the four aspects, referring to these experiences as realms having different combinations of passive–active participation and absorption–immersion experiences (Fig. 14.1).

First as a form of entertainment, an experience implies that the traveller's participation is rather passive and that the experience 'goes into the customer' (Mossberg, 2015). Second the traveller's connection to an event is considered an absorptive experience. An absorptive connection to an event is also suggested for educational experiences, such as culinary tours, in which the traveller is expected to be active. Third escapist experiences, such as a cooking class, involve active participation and immersion, and the customer 'goes into the experience' (Mossberg,

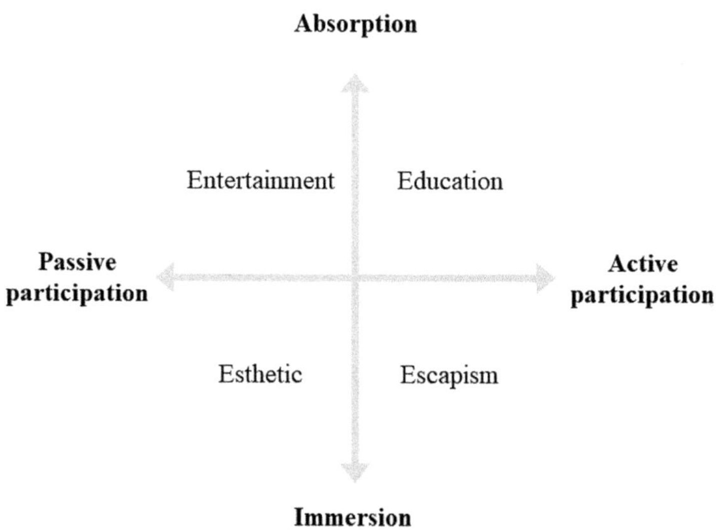

Figure 14.1 The four dimensions of an experience.
Source: Modified from Pine, J., & Gilmore, J. (1999). *The experience economy* (p. 30). Boston, MA: Harvard Business School Press.

2015). Fourth esthetic experience implies a passive participation though the traveller may be immersed in the event.

Memorable experiences and those that affect behavior encompass all the aspects of the four realms (Stone, Soulard, Migacz, & Wolf, 2018). It is also noteworthy that travellers are part of an event and serve as one of the cues affecting other travellers' experiences even if they play a passive role at the event (Shonk & Chelladurai, 2008).

14.2 Methods

A quantitative research approach was applied for empirical analysis. In addition, this study employed a survey research method using a self-administered questionnaire.

14.2.1 Data collection and analysis

Data used for this study were collected through a questionnaire distributed to people visiting the MATKA Nordic Travel Fair, an annual travel fair in Helsinki, Finland (https://matka.messukeskus.com). This fair attracts visitors of different ages and genders from all over Finland. People visit the travel fair to obtain the latest information, inspiration and special offers. The authors of this study are interested in Finnish tourists' travel behavior and food experiences within tourism. Thus the fair was considered a suitable place to reach respondents who are in a mood to travel and are looking for travel information (Björk & Kauppinen-Räisänen, 2017a).

The questionnaire used in this study was developed and tested for validity and reliability by Björk and Kauppinen-Räisänen (2016a, 2016b). In this chapter, the study takes a case study approach in which the unit of analysis is the Finnish traveller (e.g., Yin, 2009), which here is considered as the case. The results from two sections of a lengthy questionnaire are reported in this chapter. The first section was used to measure the respondents' demographic profile (gender, age, marital status, and educational level), travel behavior (in terms of number of annual national and international holiday trips), travel companion and perceived travel experience ('How experienced are you as a traveller?'). Two questions were asked to monitor the respondents' general interest in food and determine whether food is a motive for travelling. In addition, two questions were also used to measure whether expected good and different food experiences would influence the respondent's destination choice. Finally one question was asked to determine whether food and eating at travel destinations add value to the traveller's travel experiences. For these questions, the respondents were given a four-point Likert scale (from 1 = not at all to 4 = extremely) from which to gauge their answers (Björk & Kauppinen-Räisänen, 2017b).

The second section of the questionnaire showed different items that may influence the respondents' food experiences at destinations. Their answers are provided on a four-point Likert scale (from 1 = not at all to 4 = extremely). Eleven items specified as food events were used to measure the influence of different eating encounters of two main categories, namely, immersive and absorptive. Six food and eating encounters were defined as immersive as they are characterized by presence and doing, while five encounters were defined as absorptive as they are characterized by feeling and learning (Mossberg, 2015).

The sample profile, that is, the demographics of the respondents, was identified using frequency analysis. An independent sample *t*-test, correlation analysis, and analysis of variances were conducted to examine the significant differences in factors that influence food experiences at destinations with a focus on the relevance of the basic senses in immersive and absorptive experiences.

14.3 Findings

14.3.1 Sample profile

The respondents' demographic profile portrays a sample of middle-aged individuals (mean age 45 years) with travel histories matching that of the Finnish population, that is, two national and two international vacation trips annually (Statistics Finland) (Table 14.1).

Table 14.1 Demographic profile of the sample.

Sample profile ($n = 243$)		
Gender	Women	200 (82.3%)
	Men	43 (17.7%)
Age	Mean age	45.68 ($s = 15.17$)
Social status	Unmarried	75 (31.5%)
	Married	136 (57.1%)
	Other	27 (11.3%)
Education	Elementary school	21 (8.7%)
	High school/vocational	122 (50.4%)
	University	99 (40.9%)
Number of annual domestic leisure trips	1	34 (14.0%)
	2	122 (50.2%)
	3	52 (21.4%)
	4	35 (14.4%)
Number of annual international leisure trips	1	16 (6.6%)
	2	163 (67.1%)
	3	52 (21.4%)
	4	12 (4.9%)

Furthermore the sample matches the visitor characteristics of the MATKA Nordic Travel Fair (2017). However, compared with Statistics Finland, the data are partially biased as women are overrepresented in the sample (Statistic Finland, 2020).

The results presented in Table 14.2 document the importance of food and travel experiences for Finnish travellers.

At a destination, food can be experienced in an active-immersive manner as the tourist is participating in the consumption processes or in a passive-absorptive way by experiencing the food at a distance. Nevertheless, good food and eating experiences, whether active or passive, emerge in immersion and absorption processes in various destinations' food events, as shown in Table 14.3. Furthermore immersive ($P = .000$) and absorptive ($P = .000$) food encounters add positively to overall travel satisfaction.

Table 14.2 Food as a motive for travelling and experiences.

Food as a motive for travelling and experiences		
	Mean	**Std. dev.**
How important is food for the success of your trip?	2.56	0.781
How important is food as a motive for travelling?	2.73	0.834
How important are good food experiences for your destination choice?	2.58	0.816
How important are different food experiences for your destination choice?	2.48	0.768
Do food experiences add value to travel experiences?	3.05	0.723

Table 14.3 Places where good food and eating experiences emerge on a destination.

Places where good food and eating experiences emerge		
Immersion	**Mean**	**Std. dev.**
Hotel restaurants	2.23	0.837
Market squares	2.38	0.818
Home of locals	2.15	0.958
Streets	2.24	0.883
Food markets	2.14	0.844
Hiking tours	2.25	0.848
Absorption		
Food and ingredients sold at markets	2.28	0.826
Food and ingredients sold on street	2.22	0.835
Food and ingredients sold in stores	2.43	0.769
Food and ingredients sold at farms	2.27	0.866
Food and ingredients sold alongside roads in the countryside	1.96	0.874

The importance of the basic senses for food experiences at destinations and the figures presented in Table 14.4 provide evidence that food experiences are multi-sensorial experiences. However, the Pearson correlation analysis shows that "taste" does not significantly add value to food and eating experiences in all encounters (values in italics represents nonsignificant).

Based on the traveller's level of interest in food (motive for travelling), they were grouped into three categories. First the findings show that some tourists (referred to as 'foodies') find destination food experiences to be a very strong motive for traveling. These foodies are essentially driven by the destination's food culture and local cuisine. Second the findings imply that other travellers (referred to as 'casual foodies') are moderately interested in food. Third a percentage of the travellers (referred as 'survivors') do not pay much attention to food. They eat mainly to boost their energy. Finally the findings of a one-way analysis of variance revealed that foodies find greater pleasure in all types of food events—immersive and absorptive—and that there is also a significant difference between the casual foodies and survivors based on compared mean differences (Table 14.5).

The three groups of travellers were also analyzed for differences in demographic characteristics and travel behavior, but no significant differences are to be reported.

14.4 Discussion and conclusion

This study analyses various food encounters that are valuable for Finnish travellers using a self-administered questionnaire distributed at the biggest tourism fair in the Nordic countries. The study focuses on destinations' foodscapes, which comprise various types of food events. These events can be categorized based on the environment, whether organized or nonorganized, and the participants, whether travellers or residents. The events are established to allow travellers to gain experiences by staying passive and absorbing information or participating in the events in a multi-sensory and immersive manner.

This study aims to explore the various roles of travellers at a destination's food events and the importance of the basic human senses for travellers' food experiences. First the empirical findings presented in this study validate previous studies claiming that destinations' food trigger travel motivation (e.g., Boniface, 2003; Hall & Sharples, 2003; Hsu et al., 2009; Kivela & Crotts, 2005). The findings also support past studies in that travellers have a great interest in positive experiences and that they desire change from their normal routine during their travels (e.g., Richards, 2002). Hence, the study found that expected good and different food experiences affect destination choice (e.g., Boniface, 2003; Hall & Sharples, 2003; Hsu et al., 2009; Kivela & Crotts, 2005). Furthermore the findings also show that food experiences contribute to travellers' holistic travel satisfaction (e.g., Kim et al., 2011; Neild et al., 2000).

Second the findings show that experiences are gained through absorption of and immersion in destinations' food events. The findings show evidence that absorptive

Table 14.4 Food encounters and the basic senses.

	Senses for food experiences							
	Smell		Vision		Touch		Taste	
	Mean	**Std. dev.**	**Mean**	**Std. dev.**	**Mean**	**Std. dev.**	**Mean**	**Std. dev.**
	2.86	0.781	2.95	0.781	2.37	0.925	3.36	0.764
Correlations								
Immersive events								
When eating in hotel restaurants	0.324[a]		0.351[a]		0.311[a]		0.153[b]	
When eating at market squares	0.349[a]		0.242[a]		0.386[a]		0.306[a]	
When eating in the home of locals	0.201[a]		0.139[b]		0.378[a]		*0.081*	
When eating street foods	0.214[a]		0.155[b]		0.251[a]		0.187[a]	
When eating at food markets	0.208[a]		0.141[b]		0.290[a]		*0.097*	
When eating on hiking tours	0.329[a]		0.171[a]		0.321[a]		0.146[b]	
Absorptive events								
Food and ingredients sold at markets	0.298[a]		0.180[a]		0.303[a]		0.165[b]	
Food and ingredients sold on streets	0.268[a]		0.206[a]		0.318[a]		0.180[a]	
Food and ingredients sold in stores	0.271[a]		0.291[a]		0.259[a]		0.252[a]	
Food and ingredients sold at farms	0.312[a]		0.255[a]		0.382[a]		*0.131*	
Food and ingredients sold alongside roads in the countryside	0.234[a]		0.135[b]		0.330[a]		*0.049*	

[a]Significant at the .01 level.
[b]Significant at the .05 level.

Table 14.5 Food events that interests different groups of travellers.

Two types of food events	Three groups of travellers					
Immersive events	**Survivors**		**Causal foodies**		**Foodies**	
	Mean	**Std. dev.**	**Mean**	**Std. dev.**	**Mean**	**Std. dev.**
	1.967	0.531	2.254	0.629	2.831	0.626
Survivors			0.286[a]		0.664[a]	
Causal foodies					0.377[a]	
Absorptive events	Survivors		Causal foodies		Foodies	
	Mean	Std. dev.	Mean	Std. dev.	Mean	Std. dev.
	1.932	0.551	2.284	0.666	2.686	0.728
Survivors			0.351[a]		0.758[a]	
Causal foodies					0.402[a]	

[a]Significant at the .01 level.

and immersive food encounters add value to travellers' food experiences. Using the framework of Pine and Gilmor (1999), the present study holds that various food events invite the travellers to either actively participate in the cocreation of food experiences, that is, to become immersed in the event, or passively reflect on the event through absorption.

Third the findings assert that food experiences emerge from multisensory emotional stimulation. Therefore all the basic senses are relevant to a food experience. However, destinations are composed of food events where other basic senses are more important than taste. For example, taste was not significantly important when eating street food and food in marketplaces and when the traveller is watching how foods and ingredients are sold at farms on the countryside. Thus the finding supports the notion that food experiences are multidimensional, where stimulation is derived through multiple senses, not only through taste.

14.5 Managerial implications

To conclude, the study implies that destination marketing organizations (DMOs) should recognize that food experiences are universal and cannot be explained through demographic variables only. Furthermore destinations' foodscapes can and should combine different types of food events to signal to DMOs that a destination's foodscape deserves to be perceived from a holistic perspective. For company and food event managers, the findings give reason to focus on the assortment of food experiences offered and flexibility in the service offering. To be able to meet the expectations of travellers who want to enjoy the food experience in either an absorptive or immersive mode, insight into customer segmentation is needed. Not only that but also the service processes of the food events have to be in their interaction reflexive and adaptable in case, during the process, a customer switches from an absorptive to an immersive mode or vice versa.

References

Björk, P., & Kauppinen-Räisänen, H. (2016a). Exploring the multi-dimensionality of travellers' culinary gastronomic experiences. *Current Issues in Tourism*, *19*(12), 1260–1280.

Björk, P., & Kauppinen-Räisänen, H. (2016b). Local food: A source for destination attraction. *International Journal of Contemporary Hospitality Management*, *28*(1), 177–194.

Björk, P., & Kauppinen-Räisänen, H. (2017a). Destination's gastronomy for holiday well-being. *British Food Journal*, *119*(7), 1578–1591.

Björk, P., & Kauppinen-Räisänen, H. (2017b). Interested in eating and drinking? How food affects travel satisfaction and the overall holiday experience. *Scandinavian Journal of Tourism and Hospitality*, *17*(1), 9–26.

Björk, P., & Kauppinen-Räisänen, H. (2019). Destination foodscape: A stage for travellers' food experience. *Tourism Management*, *71*(April), 466–475.

Boniface, P. (2003). *Tasting tourism: Traveling for food and drink*. Burlington, VT: Ashgate Publishing.

Carù, A., & Cova, B. (2003). Revisiting consumption experience: A more humble but complete view of the concept. *Marketing Theory*, *3*(2), 267–286.

Cummins, S., & Macintyre, S. (2002). A systematic study of an urban foodscape: The price and availability of food in Greater Glasgow. *Urban Studies*, *39*(11), 2115–2130.

de Jong, A., & Varley, P. (2017). Food tourism policy: Deconstructing boundaries of taste and class. *Tourism Management*, *60*, 212–222.

Frisby, W., & Getz, D. (1989). Festival management: A case study perspective. *Journal of Travel Research*, *28*(1), 7–11.

Getz, D. (1997). *Event management and event tourism*. New York: Cognizant Communication.

Hall, M., & Sharples, L. (2003). The consumption of experiences or the experience of consumption? An introduction to the tourism of taste. In M. Hall, L. Sharples, R. Mitchell, N. Macinois, & B. Cambourne (Eds.), *Food tourism around the world: Development, management and markets* (pp. 1–24). Oxford: Butterworth-Heinemann.

Hegarty, J., & O'Mahony, B. (2001). Gastronomy: A phenomenon of cultural expressionism and an aesthetic for living. *Hospitality Management*, *20*(1), 3–13.

Hjalager, A., & Richards, G. (2002). *Tourism and gastronomy*. London: Routledge.

Hsu, T.-K., Tsai, Y.-F., & Wu, H.-H. (2009). The preference analysis for tourist choice of destination: A case study of Taiwan. *Tourism Management*, *30*(2), 288–297. Available from https://finland.mfa.gov.ua/storage/app/sites/28/imported_content/5df2c911f0ef3.pdf.

Kauppinen-Räisänen, H., Gummerus, J., & Lehtola, K. (2013). Eating experiences described by self, food, place, time, and scape. *British Food Journal*, *115*(5), 666–685.

Kim, Y. H., Kim, M., Goh, K. B., & Antun, M. J. (2011). The role of money: The impact on food tourists' satisfaction and intention to revisit food events. *Journal of Culinary Science & Technology*, *9*(2), 85–98.

Kivela, J., & Crotts, J. C. (2005). Gastronomy tourism: A meaningful travel market segment. *Journal of Culinary Science & Technology*, *4*(2/3), 39–55.

Long, L. (2010). Culinary tourism and the emergence of an Appalachian cuisine: Exploring the "foodscape" of Asheville, NC. *Folklore Journal*, *51*(1), 4–20.

Mason, N. C., & Paggiaro, A. (2012). Investigating the role of festivalscape in culinary tourism: The case of food and wine events. *Tourism Management*, *33*(6), 1329–1336.

Mikkelsen, B. (2011). Images of foodscape. Introduction to foodscape studies and their application in the study of healthy eating out-of-home environments. *Perspectives in Public Health*, *131*(5), 209–216.

Mossberg, l (2015). *Att skapa upplevelser – Från OK till WOW*. Lund: Studentlitteratur.

Neild, K., Kozak, M., & LeGrys, G. (2000). The role of food service in tourist satisfaction. *International Journal of Hospitality Management*, *19*(4), 375–384.

Okumus, B., Okumus, F., & McKercher, B. (2007). Incorporating local and international cuisines in marketing tourism destinations: The cases of Hong Kong and Turkey. *Tourism Management*, *28*(1), 253–261.

Organ, K., Koenig-Lewis, N., Palmer, A., & Probert, J. (2015). Festivals as agents for behaviour change: A study of food festival engagement and subsequent food choices. *Tourism Management*, *48*, 84–99.

Pine, J., & Gilmore, J. (1999). *The experience economy*. Boston, MA: Harvard Business School Press.

Quadri-Felitti, D., & Fiore, A. M. (2012). Experience economy constructs as a framewoek for understanding wine tourism. *Journal of Vacation Marketing*, *18*(1), 3–15.

Quan, S., & Wang, N. (2004). Towards a structural model of the touristic experience. *An illustration from food experience in tourism. Tourism Management, 25*(3), 297–305.

Richards, G. (2002). Gastronomy: An essential ingredient in tourism production and consumption? In A.-M. Hjalager, & G. Richards (Eds.), *Tourism and gastronomy* (pp. 3–20). London: Routledge.

Ryu, K. (2005). *DINESCAPE, emotions, and behavioral intentions in upscale restaurants* (Unpublished doctoral dissertation). Manhattan, KS: Kansas State University.

Santich, B. (2004). The study of gastronomy and its relevance to hospitality education and training. *Hospitality Management, 23*(1), 15–24.

Shonk, D., & Chelladurai, P. (2008) Service quality, satisfaction, and intention to return in event sport tourism. *Journal of Sport Management, 22*, 587–602.

Sims, R. (2009). Food, place and authenticity: Local food and the sustainable tourism experience. *Journal of Sustainable Tourism, 17*(3), 321–336.

Sobal, J., & Wansink, B. (2007). Kitchenscapes, tablescapes, platescapes, and foodscapes: Influences of microscale built environments on food intake. *Environment and Behavior, 39*(1), 124–142.

Spence, C. (2018). *Gastrophysics. The new science of eating*. London: Penguin Books Ltd.

Statistic Finland. (2020). Population / Population structure (accessed 12.02.20).<http://pxnet2.stat.fi/PXWeb/pxweb/en/StatFin/>.

Stone, M., Soulard, J., Migacz, S., & Wolf, E. (2018). Elements of memorable food, drink, and culinary tourism experiences. *Journal of Travel Research, 57*(8), 1121–1132.

Yin, R. K. (2009). *Case study research: Design and methods* (4th ed.) Thousand Oaks, CA: Sage. Trudie Aberdeen University of Alberta.

Creating food experiences: summary and implications for practitioners

15

Adrienne Steffen and Susanne Doppler
University of Applied Sciences Fresenius Heidelberg, Heidelberg, Germany

15.1 Summary of findings

Due to changing market and competitive conditions, a purely transaction-oriented marketing is becoming less important (Bruhn & Hadwich, 2012, p. 23). Instead, firms are using customer experiences to differentiate themselves from competitors. Customer experience management (CEM) is a process of strategically managing all customer experiences with a provider at all customer touchpoints (Schmitt, 2003, p. 17).

The objective is to optimally design all experiences of a customer at all customer touchpoints of a provider. This strategic management process puts the creation of unique and extraordinary experiences for the customer in the foreground (Bruhn & Hadwich, 2012, p. 23), a practice increasingly used by the food industry. The remainder of this chapter presents the main findings from all chapters and discusses practical implications.

15.1.1 Summary of the findings on the theoretical background of food experiences

The first section of the book reviewed experience management background literature and laid the groundwork for understanding food experiences.

Chapter 1, Introduction to the Experience Economy, by Doppler and Steffen briefly introduced the history of CEM. The idea of hedonic consumption arose in the early 1980s when Hirschman and Holbrook (1982) wrote a paper on hedonic consumption. Since then the creation of a customer experience and strategically managing the customer experience has gained increasing attention. This book takes a holistic view on the customer experience and defines it as: "the sum of all impressions a customer has with a company over the span of his customer lifetime including individual perceptions, interactions and the quality of the company performance" (freely translated from Bruhn & Hadwich, 2012, p. 10). To create a customer experience, firms want to manage the customer journey efficiently across all touchpoints to allocate resources.

Case Studies on Food Experiences in Marketing, Retail, and Events. DOI: https://doi.org/10.1016/B978-0-12-817792-1.00015-0

Pohlman provides an overview of general consumer choice models such as the stimulus–organism–response model (Jacoby, 2002) and the Engel–Blackwell–Miniard model (Engel, Blackwell, & Miniard, 1995) in Chapter 2, Introduction to Consumer Food Choices. He further introduces the total food quality model (Grunert, 2005) to explain and predict food consumption choices. The chapter argues that early choice models assumed food choices to be rational. Yet, further research, the Engel–Blackwell–Miniard model and the total food quality model, challenge this rational food choice focus. The chapter concludes that nonrational factors, for example, environmental factors increasingly influence food choices. The clean label trend is discussed as consumers have become interested in product credence attributes and care about the origins of the ingredients and the production methods.

When discussing the role of food in creating experiences in Chapter 3, The Role of Food in Creating Experiences, Doppler and Steffen propose that food is more than a human need. Although a few people who are stressed and short of time occasionally rely on fast food to get basic nutrition, the majority of people see food as a quality time, for example, with their family and friends with culturally embedded food habits and social norms. In sum, people see food shopping, food preparation, and food consumption as ways to satisfy higher level needs. The chapter furthermore suggests that marketers design and stage unique food experience. Therefore the experience gastronomy and event marketers evoke memories and stage a unique servicescape, which one could call experiencescape. In tourism, unique food cultures are used for destination branding.

In Chapter 4, Marketers' Unintended Consequences of Improving the Consumer Food Experience: Discovering the Key Drivers, Capriello and Hassan used a text mining tool to investigate and map the key drivers of the unintended consequences of enhancing the food experience such as brand, price, and customer experience. The unintended consequences of enhancing the food experiences included "obesity, materialistic orientation, customers' unaware decision processes, and overeating" (Chapter 4, Capriello and Hassan). The researchers concluded that a firms' strategic orientation in service design is a further key driver for consumer behavior.

15.1.2 Summary of findings on shopping for food experiences

In the second section of the book, case studies on the shopping for food experience were investigated.

In Chapter 5, Crafting Food Shopping Experiences: The Case of Food Halls in Luxury Departments Stores, Koronaki and Theodoridis show that retail outlet managers create holistic food shopping environments which combine utilitarian and hedonic elements. The research team uses the three elements of the retail experience (atmospheric, physical, and social) to examine how these elements are used to create entertaining, educational, escapist, and esthetic experiences in luxury department stores. The authors conclude that food retail outlets should understand how these elements create a holistic customer experience and that firms should build a story about their brand and their brand identity.

In Chapter 6, Organic Food Shopping in Australia and Bangladesh: The Role of Perceived Organic Food Value in the Attitude–Intention–Behavior Model, Sultan and his research team examine the attitude–intention–behavior paths of the theory of planned behavior in a cross-country study with a sample of organic food shoppers from Australia and from Bangladesh. They added perceived organic food value as an antecedent for both shopping intention and behavior. The results indicate that organic food shopping intention and shopping behavior are influenced by the consumers' shopping attitudes and their health-values.

In Chapter 7, Promoting Organic Products in India: The Case of Two Brothers Organic Farms, Kumar et al. present a case of two brothers who started an organic farm in India. The research team identified the farm's entrepreneurial success factors. The brothers' commitment, enthusiasm, and hard work paid off so that the two brothers achieved a top position in the Indian organic food market. One of the main success factors was the identification and education of potential customers. The brothers conducted many workshops on and off the farm, wrote blogs, and communicated on social media. Farm visits and farm stays were introduced to cocreate with customers and to add new products to their portfolio. In addition, the brothers used existing channels such as Amazon and farmers' markets for distribution. Especially the farmers markets contributed to the food experience because potential consumers could experience the product before buying it

15.1.3 Summary of findings on food consumption experience in restaurants

The third section of the book focuses on food consumption experiences in restaurants. In Chapter 8, Engineering the Experience in Restaurants by Means of Process Design and Management, Hanson studied business process modeling (BPM) in four top restaurants in Campo Grande, Brazil. From the case it can be learned that restaurant managers who have a professional view and clearly understand BPM will grow and profit. The case shows, particularly the one of Les Amis Bistrô and, to a certain extent, Domus Bistrot, that good food and a nice ambiance but irregular service results when management is amateurish. Both restaurant managers see the need for the implementation of BPM to grow further. The chapter further points out that experiences should be designed and carefully planned to delight the customers. The case of Imakay and 00 Pizza & Pane show how BPM can be done. All details within the process are carefully designed, for example, with using fresh ingredients, a lovely décor, a careful menu design, dish recipes, the kitchen layout, workflow and staff performance, and stock keeping to create an outstanding customer experience.

Bauer-Krösbacher et al. explain in Chapter 9, Ski Huts in Austria and their Culinary Offer: A Question of Quality vs. Quantity?, that food experiences are important for tourists on vacation. The research team investigates dining in the context of ski huts, which usually deal with more challenges due to their location on the mountain, for example, in terms of numbers of people, logistics, and staff compared with regular restaurants. The offered food at ski huts is mass production. The interviews with ski-hut operators showed that there are peak times. Yet the off-peak times can be used to offer smaller theme-based meals to improve capacity utilization and also to offer additional food experiences throughout the day. The case further showed that the use of local food further enhances the unique experience and that customers would like to have more information about these local suppliers. In addition, the findings indicate that Austrian skiing destinations still have potential to improve their communication and marketing effort to differentiate themselves from competition.

In Chapter 10, Food Experience Design for Plant-Based Restaurants, Bertella investigates possible experience design strategies of plant-based restaurants. These restaurants do not explicitly target vegetarians or vegans but can appeal to a broader market segment. The case study suggests that the vegetarian market segment is a growing market phenomenon. Besides the adherence to vegetarianism, gender, and local perception of vegetarianism could be crucial segmentation criteria when designing plant-based experience design strategies.

15.1.4 Summary of findings on food experience and events

Chapter 11, Event Catering: Enhancing Customer Satisfaction by Creating Memorable Holistic Food Experiences, presents the German case of an event caterer. The authors Doppler, Steffen, and Wurzer show in the case that event catering plays a central role for the quality of experience and customer satisfaction during the event. The caterer should define its individual values as a guideline for the service so that memorable, holistic food experiences can be created. The sensory dimensions seem to be particularly important in the event catering context. When it comes to the implementation, the quality of the food and ingredients, the service quality in the preevent and live-event phase, the presentation of the food and the harmony between event design, atmosphere, and food are decisive factors for a memorable holistic food experience.

In Chapter 12, "We Have Our style!" Optimizing Food Experiences in Personalized Catering Boutique Concepts, Velasco Vizcaino et al. investigated experiential aspects of luxury catering services, in particular restaurant innovativeness and the structure of luxury catering experiences. Critical incidents were identified by interviewing managers and frontline employees. It was shown that boutique catering is a complex process. Successful boutique caterers understand this complexity and use a lot of tangible and experiential resources to create a unique, high-quality, and holistic experience. The results indicate that luxury caterers can get a competitive advantage when: "the firm communicates the difference between service and servility, when they integrate and contextualize new information, when the firm facilitates social interaction and lastly, when the firm implements entrepreneurial marketing approaches to foster innovative capacities" (Chapter 12, Velasco Vizcaino et al).

In Chapter 13, Curated Nostalgia—Constructing and Consuming the City Through a Culinary Adventure, Stovall et al. examined Los Angeles Eats Itself, which is a curated, scripted dinner series. The dinner series creates an experience by relating to the city's historical past and its identity. Visitors could experience the past through memories of others and the food was related to the specific time period. According to the authors, nostalgia was particularly helpful in assisting in storytelling and in the cocreation of experiences.

In Chapter 14, Experiences Gained Through Absorption of and Immersion in Destinations' Food Events—Case Evidence from Finnish Travellers, Björk and Kauppinen-Räisänen show that travelers' experiences are gained through absorption and immersion in destinations' food events. The study which is conducted in Finland using questionnaires at the MATKA Nordic Travel Fair focuses on foodscapes at various types of events. The results indicate that food experiences are multisensory and that all senses are relevant in building the experience, not only taste. These food experiences influence travel motivation and travel satisfaction.

15.1.5 Lessons learned for practitioners

The following section summarizes the lessons learned from all cases. All recommendations made from all authors were therefore collected, summarized, and categorized. Fig. 15.1 shows the four main fields of recommendations which emerged from the order process. Table 15.1 then further shows the summary of recommendations for practitioners which emerge in the four categories.

15.1.5.1 Information needs

Chapters 2, 5, 12, and 14 addressed the role of firm's information needs in their recommendations. Pohlmann shows in Chapter 2, Introduction to Consumer Food Choices, that nonrational factors heavily influence food choices, often on an unconscious level. Food choices are made repetitively so that consumers form consumption habits and take mental shortcuts to make quicker decisions. These habits are a challenge for the individual as well as for the marketer. Once these habits have been formed, the individual has problems when wanting to change his or her diet. Even the marketer who wants to change an individual's behavior will get consumer resistance due to his or her formed habits. Yet the chapter indicates that consumers are becoming more interested about the food's origin and production methods and the resulting conflicts between marketers and consumers offer opportunities for both parties. It can therefore be suggested that marketers and retailers need to understand consumption behavior and particular consumption habits which are difficult to break. Based on market research and the resulting knowledge of their

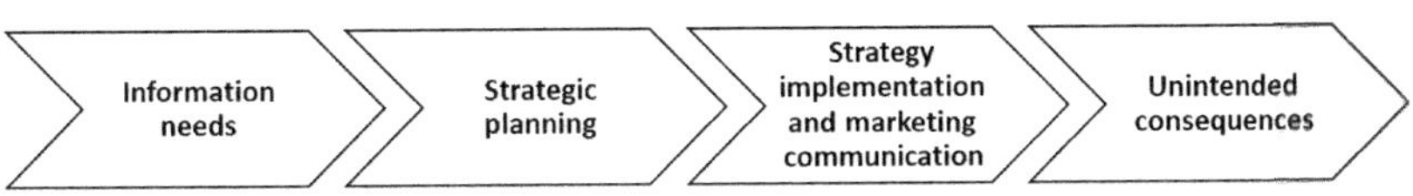

Figure 15.1 The emerging four fields for recommendations.

Table 15.1 Summary of recommendations for practitioners.

Information needs
• Marketers and retailers need to know about consumption behavior (Pohlmann, Chapter 2: Introduction to Consumer Food Choices). • The firms should understand the value customers seek for their visits in the outlets (Koronaki and Theodoridis, Chapter 5: Crafting Food Shopping Experiences: The Case of Food Halls in Luxury Departments Stores). • Caterers should listen carefully to the wishes and emotional needs of their customers and have to be able to translate these to successful service experiences (Velasco Vizcaino et al., Chapter 12: "We Have Our style!" Optimizing Food Experiences in Personalized Catering Boutique Concepts). • The retail outlets should get an understanding how various experiences can be designed with store elements (Koronaki and Theodoridis, Chapter 5: Crafting Food Shopping Experiences: The Case of Food Halls in Luxury Departments Stores). • Insight into customer segmentation is needed to be able to meet the expectations of travelers who want to enjoy the food experience in either an absorptive or an immersive mode (Björk and Kauppinen-Räisänen, Chapter 14: Experiences Gained Through Absorption of and Immersion in Destinations' Food Events—Case Evidence from Finnish Travellers).
Strategic planning
Developing a competitive advantage • That smart exploitation of existing distribution channels can enhance the customer experience at entrepreneurial stage (Kumar et al., Chapter 7: Promoting Organic Products in India: The Case of Two Brothers Organic Farms). • In short- and long-term planning, an adaptation to the changing environment is necessary for a good company performance (Bauer-Krösbacher et al., Chapter 9: Ski Huts in Austria and their Culinary Offer: A Question of Quality vs. Quantity?). • Develop core competence profile (Doppler and Steffen, Chapter 1: Introduction to the Experience Economy). • Extend the experience to multiple stakeholders (Kumar et al., Chapter 7: Promoting Organic Products in India: The Case of Two Brothers Organic Farms). • Differentiation is possible by offering regional and seasonal products to become regional ambassadors (Bauer Krösbacher et al., Chapter 9: Ski Huts in Austria and their Culinary Offer: A Question of Quality vs. Quantity?). • Staff recruitment and training is important for offering food quality and service in plant-based restaurants and for building a competitive advantage (Bertella, Chapter 10: Food Experience Design for Plant-Based Restaurants). • Communication and self-presentation is of utmost importance in the planning process and also during the event (Velasco Vizcaino et al., Chapter 12: "We Have Our style!" Optimizing Food Experiences in Personalized Catering Boutique Concepts). • Cateres should look for strategic partners, for example, location providers, décor, entertainment, and lighting to provide extraordinary customer experiences (Velasco Vizcaino et al., Chapter 12: "We Have Our style!" Optimizing Food Experiences in Personalized Catering Boutique Concepts).

(*Continued*)

Table 15.1 (Continued)

- Using narratives can help tremendously in creating a unique event concept (Stovall et al., Chapter 13: Curated Nostalgia—Constructing and Consuming the City Through a Culinary Adventure).
- Destinations' foodscapes can and should combine different types of food events to signal to destination marketing organizations that a destination's foodscape deserves to be perceived from a holistic perspective. Companies and food event managers should focus on the assortment of food experiences offered and flexibility in the service offering (Björk and kauppinen-Räisänen, Chapter 14: Experiences Gained Through Absorption of and Immersion in Destinations' Food Events—Case Evidence from Finnish Travellers).
- The interaction in service processes of the food events have to be reflexive and adaptable in case, during the process, a customer switches from an absorptive to an immersive mode or vice versa (Björk and Kauppinen Räisänen, Chapter 14: Experiences Gained Through Absorption of and Immersion in Destinations' Food Events—Case Evidence from Finnish Travellers).

Business process modeling

- Transparency and a strategic planning process are of utmost important (Doppler et al., Chapter 11: Event Catering: Enhancing Customer Satisfaction by Creating Memorable Holistic Food Experiences).
- Redesign their business processes and core products with technical innovations to offer a sustainable business model and a unique experience (Doppler and Steffen, Chapter 1: Introduction to the Experience Economy).
- Business process modeling and excellent management can bring restaurants to an extraordinary level of service and overall customer experience (Hanson, Chapter 8: Engineering the Experience in Restaurants by Means of Process Design and Management).

Strategy implementation

- Retailers and marketers need to inform customers about food origin and production methods (Pohlmann, Chapter 2: Introduction to Consumer Food Choices).
- A brand should be developed, for example, the firms should choose which experience dimensions to build on, develop a story around the brand, and create a brand identity (Koronaki and Theodoridis, Chapter 5: Crafting Food Shopping Experiences: The Case of Food Halls in Luxury Departments Stores).
- Firms should build and enhance consumer attitude for and perceived values toward the organic food items by considering costs, claims, certifications, and communication (Sultan et al., Chapter 6: Organic Food Shopping in Australia and Bangladesh: The Role of Perceived Organic Food Value in the Attitude–Intention–Behavior Model).
- Communication channels such as blogs and YouTube help firms in their entrepreneurial stage when they have little communication budget (Kumar et al., Chapter 7: Promoting Organic Products in India: The Case of Two Brothers Organic Farms).
- Plant-based restaurants should communicate the nutritional aspects of food and have a theme relevant to local culture to create successful customer experiences (Bertella, Chapter 10: Food Experience Design for Plant-Based Restaurants).

Unintended consequences

- The unintended consequences were found to be obesity, a consumer's materialistic orientation, consumer overeating, and customer' unaware decision processes (Capriello and Hassan, Chapter 4: Marketers' Unintended Consequences of Improving the Consumer Food Experience: Discovering the Key Drivers).

motivations and consumption habits they are likely to meet the demands of the critical consumer.

Similarly, Koronaki, and Theodoridis propose, in Chapter 5, Crafting Food Shopping Experiences: The Case of Food Halls in Luxury Departments Stores, that luxury, premium, and mainstream food outlets should get an understanding of how various experiences can be designed with store elements. Furthermore the firms should understand the value customers seek for their visits in the outlets.

The research of Velasco Vizcaino et al. suggests in Chapter 12, "We Have Our Style!" Optimizing Food Experiences in Personalized Catering Boutique Concepts, that caterers should listen carefully to the wishes and emotional needs of their customers and have to be able to translate these to successful service experiences.

Björk and Kauppinen-Räisänen also put forward in Chapter 14, Experiences Gained Through Absorption of and Immersion in Destinations' Food Events—Case Evidence from Finnish Travellers, that a firm needs customer insights for an efficient customer segmentation to be able to meet the expectations of travelers who want to enjoy the food experience in either an absorptive or immersive mode.

The chapters thus suggest that firms should research their customer needs and food habits carefully. Only with enough information they can make informed decisions, for example, for product design or to take a customer offer in the catering business. Once they have collected enough information, they can start the strategic planning process.

15.1.5.2 Strategic planning

Several chapters addressed the importance of the firm's strategy for CEM in food industries. Some chapters discussed how firms can develop a competitive advantage with their customer experience, others how BPM helps to create an extraordinary customer food experience.

15.1.5.2.1 Developing a competitive advantage

Chapter 1, Introduction to the Experience Economy, by Doppler and Steffen suggests that firms generally need to change the way they conduct business and that they should think about their future competence profile to create and deliver unique experiences in the future. Based on the idea of Lemon and Verhoef (2016), firms should integrate several business functions such as information technology, logistics, marketing, service operations, human resources, and external partners.

Kumar et al. showed in Chapter 7, Promoting Organic Products in India: The Case of Two Brothers Organic Farms, that Two Brothers Organic Farm enhanced the customer experience by developing a competitive advantage. First, they exploited existing distribution channels to enhance the customer experience. Second, they got customers to the farm through farm visits and farm stays. This practice enriched the customer experience and customers became part of the value creation process. Third, the competitive advantage was built by organizing workshops on organic farming and events and harvest time were used to extend the experience to multiple stakeholders, for example, people who plan on starting their own organic farm or entrepreneurship-trainers to motivate their entrepreneurs.

Bauer-Krösbacher et al. show in Chapter 9, Ski Huts in Austria and their Culinary Offer: A Question of Quality vs. Quantity?, that mass production does not need to compromise on quality. Ski-hut operators who have extreme peeks and enormous demands, can offer high-quality dishes and can differentiate themselves from competitors this way. Practitioners can learn for their short- and long-term planning that an adaptation to the changing environment is necessary for such a good performance. Austrian ski-hut operators focus on regional and seasonal products to differentiate themselves from competitors and to become "ambassadors of the Austrian cuisine" (Chapter 12, Bauer-Krösbacher et al).

Bertella suggests in Chapter 10, Food Experience Design for Plant-Based Restaurants, that food experiences in regular and plant-based restaurants have similar drivers, for example, an elegant but at the same time environmentally friendly interior. Yet it seems as if plant-based restaurants which pay attention to nutritional aspects of food and which have a theme that is not in conflict with local cultures are particularly successful in creating customer experiences. Plant-based restaurants should recruit and train their staff carefully especially concerning food quality and service. This knowledge advantage seems crucial for building a competitive advantage in a market which is just forming and growing quickly.

The research of Velasco Vizcaino et al. in Chapter 12, "We Have Our Style!" Optimizing Food Experiences in Personalized Catering Boutique Concepts, proposes that communication and self-presentation is of utmost importance in the planning process and also during the event. A recommendation by the research team was made to decline requests if the potential for conflicting interests is too high. In addition, the researchers stress the importance of building long-term strategic partnerships.

The challenge for caterers is to transform ideas arising from the customer's personal history into a professional catering concept and extraordinary experiences which then generate word of mouth and social media presence.

A further example for developing a competitive advantage in the event context is suggested by Stovall et al. The authors suggest in the case study *Los Angeles Eats Itself* that "narrativity is as helpful in understanding our consumption of the city as it is in articulating our phenomenological and semiotic engagements with it" (Chapter 13, Stovall et al). The dinner series recreates an d reconsumes ignominious Los Angeles stories, which is a point of differentiation compared with other events. The city went beyond sensationalization of dark city episodes but went to create an event which helps to reexamine the relationship with the city's identity and the way the city is consumed. The case nicely shows how cities can bring dark historical episodes, which shape the cities' identity, and translates them into exclusive food events and experiences.

The authors recommend other cities to follow Los Angeles' road map and to also create their own, curated dinner series.

Björk and Kauppinen-Räisänen, Chapter 14, Experiences Gained Through Absorption of and Immersion in Destinations' Food Events—Case Evidence from

Finnish Travellers, suggest that destinations should combine different types of food events and that a destination's foodscape deserves to be perceived from a holistic perspective. They do not directly address the competitive advantage but suggest that companies and food event managers should focus on the assortment of food experiences offered and flexibility in the service offering, which implies that this focus leads to the generation of a competitive advantage. In addition, the service processes of food events should be reflexive and adaptable because customers need to be able to switch from an absorptive to an immersive mode and vice versa.

Overall it can be concluded that a future-oriented competence profile, the integration of customers in core processes, high product quality and regional food products, knowledgeable employees, good communication and self-presentation, and a unique, story-focused business concept help to create an outstanding customer food experience, which differs from the competition.

15.1.5.3 Business process modeling

Once a firm decides on a competitive advantage, the business processes need to be modeled for a smooth operation. In addition, technical developments will change the way firms offer experiences. Many firms will have to change their core product, offer a total experience, and redesign their business processes. Virtual reality, mobile Internet, and social media changes the way people live and they are changing the way firms offer experiences. For example, rather than to visit museums or sights, customers can experience the exhibits online (Smit & Melissen, 2018, p. 2). Doppler and Steffen note in Chapter 1, Introduction to the Experience Economy, that a major challenge for firms will be to identify these process-changing experiences and design them to achieve a competitive advantage and to create a sustainable business model.

In Chapter 8, Engineering the Experience in Restaurants by Means of Process Design and Management, Hanson proposes that transparency and a strategic planning process are of utmost important. By using an example from the event catering context they propose that a professional view on management and processes is necessary and that business processes need to be clearly defined. The defined guidelines are to be used by all employees for every customer contact and will enhance customer satisfaction. The case suggests that the size of the firm does not matter. For planning events of every size, event catering should be integrated into the event conception for quality assurance. The authors note that only those restaurants will sustain longer than 5 years which have excellent management and use BPM. These restaurants can offer their customers an excellent service and overall customer experience.

The case study of Bauer-Krösbacher et al. in Chapter 9, Ski Huts in Austria and their Culinary Offer: A Question of Quality vs. Quantity?, is devoted entirely to BPM. It shows that restaurants which want to survive in Brazil in the long term, should understand BPM. A good food quality, service and the restaurant ambiance are important factors but according to the authors, only BPM and excellent management can bring the restaurant to an extraordinary level of service and overall customer experiences.

The discussed chapters emphasize the importance of BPM. The authors of Chapters 1, 8, and 9 recommend using BPM to redesign and optimize business processes. Firms should identify process-changing experiences in the time of digitalization. All modeled business processes need to be transferred into employee guidelines to offer the customer an extraordinary experience and to assure the experience quality.

15.1.5.4 Strategy implementation and marketing communication

Several authors made recommendations concerning the strategy implementation and here especially about the marketing communication. Pohlmann, for example, recommends in Chapter 2, Introduction to Consumer Food Choices, that retailers and marketers need to inform customers about food origin and production methods. More detailed recommendations concerning food retail outlets were proposed in Chapter 5, Crafting Food Shopping Experiences: The Case of Food Halls in Luxury Departments Stores, by Koronaki and Theodoridis. The authors propose that luxury food outlets should get an understanding of how various experiences can be designed with store elements in a first step. In a second step, the brand should be developed, for example, the firms should choose which experience dimensions to build on, develop a story around the brand and create a brand identity. Furthermore the firms should understand the value customers seek for their visits in the outlets. The chapter therefore combines information needs in the strategy design with its implementation and its communication to customers.

Further recommendations regarding communication are provided by Sultan and his research team in Chapter 6, Organic Food Shopping in Australia and Bangladesh: The Role of Perceived Organic Food Value in the Attitude–Intention–Behavior Model. The researchers suggest that organic food managers and marketers can influence consumer shopping intention and behavior positively by building consumer attitude for and perceived values toward the organic food items. They proposed a 4C's strategy for how organic food managers and marketers can enhance consumer attitudes and perceived values, namely, costs, claims, certifications, and communications.

In Chapter 10, Food Experience Design for Plant-Based Restaurants, Bertella also draws attention to experience drivers which seem to be similar for regular and plant-based restaurants. The interior should for example be elegant but at the same time environmentally friendly. Yet it seems as if plant-based restaurants which pay attention to nutritious aspects of food and which have a theme that is not in conflict with local cultures are particularly successful in creating customer experiences. The chapter therefore emphasizes the role of communication.

Kumar et al. provide more detailed recommendations on how to communicate to and with customers in Chapter 7, Promoting Organic Products in India: The Case of Two Brothers Organic Farms. Especially in BRIC states like India, young entrepreneurs who have little communication budget can use communication channels like blogs and YouTube. Especially YouTube helped the recently founded farm to communicate with several stakeholders. Besides using YouTube as a free online communication, another key factor for success was to get the customers to the organic farm through farm visits and farm stays. This practice enriched the customer experience and customers became part of value creation. In addition, workshops on organic farming and events at harvest time were used to extend the experience to multiple stakeholders, for example, people who plan to start their own organic farm or entrepreneurship-trainers to motivate their entrepreneurs and to create awareness for organic farming.

To conclude, the chapters show that information to customers and honest communication to customers is a key success driver. Firms should create strong brands and build a story around their brand. When implementing the designed strategy, the Servicescape helps to enrich the customer food experience. Online and offline marketing communication is required to build consumer attitudes and to influence shopping intention. It does not matter how small the firm is, even small firms should use cheap communication tools and try to create value for the customer to experience an extraordinary food experience.

15.1.5.5 Unintended consequences

The last area for which recommendations were made is the consequences of food experience. In Chapter 4, Marketers' Unintended Consequences of Improving the Consumer Food Experience: Discovering the Key Drivers, the researchers found that the marketing strategies and policies of the firms themselves are the key drivers for the unintended consequences of improving the consumer food experience. The unintended consequences were found to be obesity, a consumer's materialistic orientation, consumer overeating, and customer' unaware decision processes.

The chapter thus suggests that firms should be aware of the consequences of their customer experiences. Firms should think critically about the impact of their customer's food experiences on a personal level and also on society as a whole.

References

Bruhn, M., & Hadwich, K. (2012). 'Customer experience – Eine Einführung in die theoretischen und praktischen Problemstellungen'. In M. Bruhn, & K. Hadwich (Eds.), *Customer experience – Forum Dienstleistungsmanagement* (pp. 3–36). Springer Gabler. Available from https://doi.org/10.1007/978-3-8349-4001-8_1.

Engel, J., Blackwell, R., & Miniard, P. W. (1995). *Consumer behavior*. Fort Worth, TX: Dryden Press.

Grunert, K. G. (2005). Food quality and safety: Consumer perception and demand. *European Review of Agricultural Economics*, *32*(3), 369–391.

Hirschman, E. C., & Holbrook, M. B. (1982). Hedonic consumption: Emerging concepts, methods and propositions. *Journal of Marketing*, *46*(Summer), 92–101. Available from https://doi.org/10.2307/1251707.

Jacoby, J. (2002). Stimulus-organism-response reconsidered: An evolutionary step in modeling (consumer) behavior. *Journal of Consumer Psychology*, *12*(1), 51–57.

Lemon, K. N., & Verhoef, P. C. (2016). Understanding customer experience throughout the customer journey. *Journal of Marketing*, *80*, 69–96. Available from https://doi.org/10.1509/jm.15.0420.

Schmitt, B. H. (2003). *Customer experience management*. Hoboken: John Wiley & Son, INC.

Smit, B., & Melissen, F. (2018). *Sustainable customer experience design – Co-creating experiences in events, tourism and hospitality*. Oxon: Routledge.

The future of food experiences

16

Susanne Doppler[1] *and Adrienne Steffen*[2]
[1]Event Management and Tourism, Hochschule Fresenius Heidelberg, Heidelberg, Germany,
[2]Consumer Behavior and Market Research, Hochschule Fresenius Heidelberg, Heidelberg, Germany

16.1 Drivers for future development

The difficulty in addressing the future is, that the future is not simply the optimized present, it can also mean to break with the status quo to provoke a change (Zukunftsinstitut, 2020, p. 50). Of course, no one knows exactly what the future will look like. Therefore it seems reasonable to elaborate why, when, and how certain incidents will occur (Yeoman, McMahon-Beattie, & Wheatley, 2015, p. 4).

Food is one of the core components of a nation's culture and identity, with its history, symbols, myths, and discourses (Bell, 1997, p. 82; Smith, 1995, as cited by Yeoman et al., 2015) and is closely linked to social relationships and cultural norms (Scott & Vallen, 2019). Consequently food cultures have changed repeatedly. Social, cultural, economic, technological, climate, and political factors were and are decisive for this change in food culture (Zukunftsinstitut, 2020, p. 12). Thus the initial point of this chapter is to consider the current social, cultural, economic, technological, climate, and political drivers of change to anticipate future developments in food consumption. Scholars point out that the society is currently in a transition from the experience economy, that came up in the 1980s (Pine & Gilmore, 1990), through the knowledge economy (Brand & Rocchi, 2011) to the transformation economy, which was already anticipated in the original studies of Pine and Gilmore. Regarding the emerging transformation economy, Brand and Rocchi (2011) argue that the future economic imperative will be built on sustainable, profitable, ethical, and fair business practices based on cooperation, value networks, and the exchange of ethical values between different stakeholders. Furthermore the authors outline a consumer mindset that is changing from self-realization to a meaningful life.

Looking toward drivers of transformation, the so-called megatrends indicate the subjects that will be shaping mankind for a long time to come. As constants of development in global society, megatrends comprise several decades. They are drivers of change, affect every single person, and encompass all levels of global society: business and politics as well as science, technology, and culture (Zukunftsinstitut, 2020, p. 14). Hence, from a general point of view, megatrends provide the background of change. Digitalization, globalization, ecologization, demographic change, politics, and socioeconomics doubtlessly cause a profound

Case Studies on Food Experiences in Marketing, Retail, and Events. DOI: https://doi.org/10.1016/B978-0-12-817792-1.00016-2

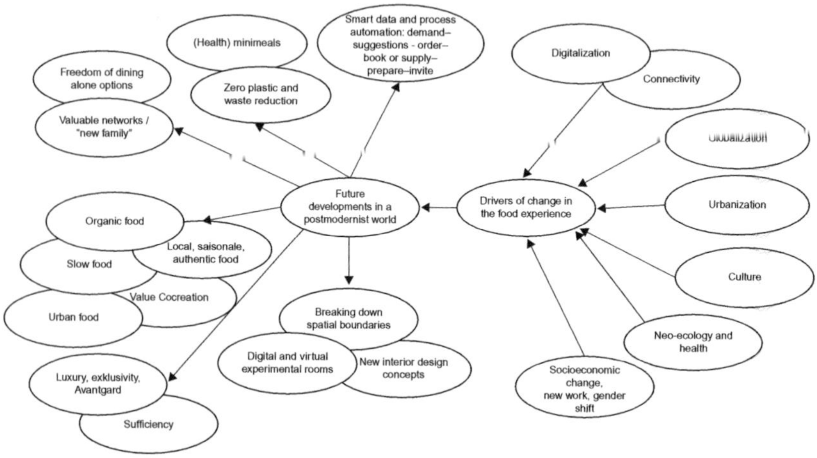

Figure 16.1 Megatrends (own presentation).

transformation worldwide and pose serious challenges to mankind, as they already have and will continue to have severe impacts on our lives (Seyler, 2020, p. 6). Since the digitalization has an impact on all relevant areas of society, such as politics, economy, science, technology, and culture, and thus leads to fundamental social changes (Zukunftsinstitut, n.d.) and it can be regarded as a central force and driver of change in general. Megatrends with a major impact on our food behavior are connectivity, urbanization, individualization, gender shift, neoecology, health, and new work (Zukunftsinstitut, 2018; as cited by Zukunftsinstitut, 2020, p. 14) (Fig. 16.1).

On the other side, food trends show lifestyles and desires, offer orientation, and provide solutions for current problems. Food trends are always carried out by people, but however, they are shaped by the profound, global, and long-term changes in megatrends (Zukunftsinstitut, n.d.). Food has become a powerful means of communication in our time (Ali, Ferdinand, & Chidzey, 2017, pp. 75–76; Zukunftsinstitut, 2020, p. 7). Food trends can therefore function as "barometers" and can help to identify developments that are affecting society at large (Zukunftsinstitut, n.d.), in our everyday eating habits they offer answers to numerous problems and needs that are triggered by social change (Zukunftsinstitut, 2020, p. 15).

16.2 Future developments of food culture

16.2.1 The postmodernist mankind

Food consumption experience plays an essential role in a postmodernist world (Yeoman et al., 2015, p. 265). In analogy to the ongoing social, economic, and technological transformation, also food consumption behavior is subject to profound changes (Casini, Contini, Romano, & Scozzafava, 2015). Hirst and Tressider (2012, p. 82) state that the phenome of rootlessness of the postindustrial, modern world brings out the relevance of diverse food experiences that meet the needs of the modern consumer, who may find himself in very different situations, places, and settings within a short period of time. For example, consumers are involved in long working days, which often relate to long-distance travel or commuting times. Consequently consumers' eating behavior shifts as well as the shopping behavior (Hirst & Tressider, 2012, p. 82–84). Hence, an equal coexistence of food consumption occurs, such as fast food consumption, consumption of finished products from retail or takeaways, restaurant visits, and online shopping with delivery at home or pick-ups at the retail market. On a business trip the consumer maybe invited to enjoy typical regional food and beverages with business partners in a restaurant. While being at home for the weekend, food experiences provide a platform for social interaction and enjoyment with families and friends (Chitakunye & Takhar, 2014).

Changing work and lifestyle choices directly influence food consumption behavior. Due to the worldwide growth of single-person households (Hay, 2015, p. 195; Statista, 2020), changes in the construct of the family as well as the development of a 365/24/7 society, substantial changes in food consumption behavior are becoming apparent (Hay, 2015, p. 194). The out of home consumption of food is increasing, mostly among single males (Casini et al., 2015). The meaning of regular three daily meals is becoming less important (Hay, 2015, p. 194; Zukunftsinstitut, 2020, p. 28–29). Instead, it becomes complemented by offers that meet the demands of individual work and lifestyle patterns, such as single dinner offers, access to food independent of time and place, the development of digital platforms that bring together friends and families scattered around the world via holograms, and enable a shared food experience independent of place (Hay, 2015). In this context, Zukunftsinstitut (2020, p. 34) sees the trend of snackification as minimeals emerging driven by the abovementioned megatrends. Defined meals dissolve into spontaneous eating opportunities that are not tied to specific times. These are eaten at home and away from home, often alone or with situational eating partners. The decision what to eat is made spontaneously, according to availability, time, and desire. This brings with it the fact that gastronomy is increasingly coming under competition from the retail trade, as retailers are also jumping on the snackification trend. To increase the quality and length of time customers spend in the stores, more and more retailers are offering gastronomy services in their stores (Lehner, 2015; Pandey & Verma, 2015). The growing demand for minimeals is leading to new types of food service concepts, supermarkets and retailers are responding with innovative "on-the-go" offers (Zukunftsinstitut, 2020, p. 38–39), for example, the supermarket chain "Rewe" in Germany. With "Rewe to go," the supermarket chain offers a diverse product range for home or immediate consumption at central locations such as train stations, gas stations, and highly frequented city locations. And that almost around the clock and anytime to go. The offer includes the categories "breakfast," "coffee to go," "cooking," "picnic," "on the road," and "with friends" (Rewe, n.d.). Also the online giant Amazon has responded to this demand for innovating "on-the-go" offers with a physical store concept. In the categories "breakfast," "lunch," and "dinner," customers can cover their needs in To Go Shops near-by (Amazon, n.d.). At the same time, delivery services with "Ready to Cook" offerings are gaining in importance (e.g., Hello Fresh, n.d.).

In addition to system gastronomy, the classic snack bars, and retailers, there are more and more lateral entrants who offer new creations of fast food with food trucks and street food stations. These similarly fast food, but often healthier, offers can cannibalize traditional snack bars and restaurants. Lateral entrants offer more flexible opening hours and the possibility of bringing the restaurant on wheels to the customer instead of the customer having to move to the restaurant and being able to drive flexibly to changing places of consumer demand. Gastronomy will have to adapt their food concepts toward healthy, vegan, and sustainable fast food as well as to portion sizes and service times.

But also the following questions come up: can fast food be personalized to individual taste? What would the perfect fast food meal look like? Coca-Cola, for example, has already installed several thousand free-style vending machines where consumers can premix and individualize their soft drink to their personal taste via app. All they have to do is scan the code at the vending machine and the product is filled in (Coca Cola, n.d.(a), n.d.(b)). This innovative idea can be transferred to the food sector. Providers and delivery services such as fast-food chains, take away restaurants, food trucks, snack bars, and even restaurants could offer both, the app-based selection of food and taste features, as well as mood- and lifestyle-related offers such as "food to relax," "food to move a mountain," "food against anger," and so on.

It is likely that the rhythm of eating three meals daily meals will get abolished in the future. Generally speaking, the gastronomy observes a shifting in the boundaries in the gastronomy, for example, between the individual courses of a menu, between the service, cook and the guests, and between the guest itself (Schwert, 2018). While for a long time the customer was considered "the king," the paradigm has shifted to considering the guest as "a friend." This development satisfies on the one hand the growing number of people living alone or eating alone during a business trip, for example, and also addresses the desire for social closeness in urban regions. This idea is confirmed by the work of Hay (2015, pp. 201–203), who has highlighted 10 trends of managing single dinners. These trends reflect two tendencies in development: on the one hand, the single diner, who likes to eat alone in an undisturbed ambience without feeling isolated, for example, in special single-person restaurants or regular restaurants with private single-person booths. On the other hand, there is also the guest who eats alone and wants to come into contact with other people who are strangers to him, to escape the feeling of "singling out," The fine balance between closeness and distance will also be reflected in restaurant concepts in the future.

16.2.2 Authenticity and local food

With regard to globalization, on the one hand an increase in international economic interrelationships can be seen, but contrary also a localization that is expressed in the strengthening of local identities and autonomies on the other hand (Yusuf, 1999). The idea of local food culture is well known, as each country and particular regions are characterized by their national and regional iconic dishes (Björk & Kauppinen-Räisänen, 2016) and there is a continuing trend in food consumption behavior, which implies an increasing consumer interest in local foods (Mirosa & Lawson, 2012). A systematic literature review on the conception and definition of local food and its systemic positioning in the agrifood production can be found in Rinaldi (2017). In summary, Martinez et al. as cited by (Rinaldi, 2017, p. 5) define local food as "food produced, processed and distributed within a geographical boundary that consumers associate with their own community." More consumers are increasingly expressing concerns about the environmental aspects of food production and consumption (Zhang, Liu, Sayogo, Picazo-Vela, & Luna-Reyes, 2016). Local foods are generally interpreted by consumers as food that is produced in a social and ecological sustainability way (Rinaldi, 2017), which helps to build trust between producers and consumers (Steffen & Doppler, 2019). Furthermore local food, integrated into alternative agricultural and food networks, "may represent a niche for rural entrepreneurs who are willing to build a closer relationship with customers" (Rinaldi, 2017, p. 5), which in turn leads to consumer satisfaction (Steffen & Doppler, 2019) and food well-being (Scott & Vallen, 2019).

In the context of local products, also the concept of authenticity has become a relevant criterion for customers buying decisions in general (Gilmore and Pine, 2007, as cited by Kim & Jang, 2016) and food in particular (Björk & Kauppinen-Räisänen, 2016). Authenticity is generally defined as "that which is believed or accepted to be genuine or real" (Taylor, 1991, p. 17, as cited by Kim & Jang, 2016), being either a staged reconstruction, for example, the ethnic restaurants in foreign countries (Kim & Jang, 2016) or reflecting a return to local food, its preparation, and presentation and the traditional food culture in a region (e.g., Aaltojärvi, Kontukoski, & Hopia, 2018; Chen & Huang, 2018; Ertugral, Cetin, Karagoz, Balık, Dincer, & Sengel, 2015; Kim & Iwashita, 2016; Lin & Mao, 2015; Privitera, Nedelcu, & Nicula, 2018; Sidali, Kastenholz, & Bianchi, 2015, Sims, 2009; Tsai, 2016). In gastronomy, poor authenticity is one of the critical factors that make restaurants, hotels and other tourism companies unsuccessful (Parsa, Self, Njite, & King, 2005). Authenticity of ethnic restaurants, for example, depends on (1) being true to itself and (2) being what it says it is, (3) the servicescape represented by the employees, (4) original dishes and taste, and (5) atmospherics related to employees (Kim & Jang, 2016). Although authentic dishes and tastes are represented via local food, which is perceived to be authentic, pure, and traditional (Kauppinen-Räisänen, Gummerus, & Lehtola, 2013), the atmospherics and the truthfulness can be reached by creating a narrative and themes. For example, the fine dining restaurant "The Fat Duck" comprises a journey the guest immediately starts once a reservation has been made (The World's 50 Best Restaurants, 2017; The Fat Duck, n.d.). This journey reflects the owner's Heston Blumenthal most significant holiday memories of his childhood, translated into a magical journey through a luxurious 15 course menu. Other examples show that authentic food experiences can be created with locally and organically produced food. In Norbert Nierkofler's three-star

Michelin restaurant in Südtirol, Italy, for example, only food organically produced in the region is used. The chef calls his concept "Cook the Mountain," with reference to the idea of supporting sustainable agriculture in the region (Beller, 2019). The result is a narrative of its own, a story about tradition, origin, heritage, and healthy food.

Zukunftsinstitut (2020, p. 18) describes this appearance of a radical shift away from the standard as the "brutal local," in which the avant-garde focuses on radical seasonality, wild food, and fermentation. The concept of the "brutal local" has emerged from an intensification, which was originally expressed as a response to globalization in the recollection of the peculiarities of regional cuisine and which has been further specified and innovatively promoted in the use of locally produced, seasonal raw materials. This also shows a deliberate avoidance of consumption of, for example, overseas products in favor of selected local products.

The slow food movement can also be seen in the context of authenticity. According to the slow food movement (slowfood.com), raw materials are used to prepare food that has been produced good, clean, and fair and that supports the sustainability of food. Slow food is prepared according to a recipe that is authentic and representative of the cultural tradition of the place and its people, where the cooking method and the equipment used (equipment) are reflected in traditional practices (Munjal, Sharma, & Menon, 2016). The movement of local food, slow food, and organic food also raises the question of the entire food supply chain, from production to consumption as a relevant part of the food experience. Information that the retailers as well as the gastronomy and caterers are required to make available to the consumer, for example, in the form of informative storytelling or with innovative digital solutions such as apps, QR codes, and Youtube channels.

Moscardo, Minihan, and O'Leary (2015, p. 215) confirm this consumer demand for knowledge in the context of food tourism experiences. They developed a conceptual framework with learning and knowledge as a core element of all food tourism experiences. The framework refers to three experiential dimensions, focusing first on place versus personal dimension, second on food versus fun dimension, and third differentiating between backstage versus frontstage experiences. The backstage focus on place refers to food themed tours and experiences, the food focus to food classes, while the frontstage focus on personal triggers fine dining experiences with iconic local chefs. The fun focus expresses itself, for example, in food festivals and food events. Food attractions and production tours combine the focus on place and fun and organic and sustainable produced food focus on personal and food (Moscardo et al., 2015, p. 215). Thus this model can also be applied for events. From an experience-oriented perspective, local food representing the place of origin, as well as local landscape and culture, become fundamental elements to create a food experience, where food quality, service quality, fun, and entertainment play an important role to create emotional experiences (Mason & Paggiaro, 2012).

But local, authentic, and organic food does not necessarily mean that production takes place in rural areas. Cities will be the preferred and predominant way of life in the future and food will consequently be produced in cities, also to avoid transport routes and to produce food at the place of demand. Microfarming and urban gardening projects already bear witness to this development (e.g., Mikrolandwirtschaft.org, n.d.). The Controlled Environment Agriculture operates in closed systems, greenhouses, or vertical farms—artificially created cultivation areas that rise up into the height, for example, along building facades—and produces plants or animals as well as products from other climate zones locally and fresh. Harvesting can take place several times a year and is unaffected by weather and climate changes and is setting the terms seasonally and locally in a new context (Zukunftsinstitut, 2020, p. 77).

From an outreach point of view it is also interesting to mention that such technical urban farms become experience spaces in cities where principles of agricultural life-cycle management and climate-friendly food production can be experienced and learned (Zukunftsinstitut, 2020, pp. 60–89) gives a good insight into current projects worldwide.

16.2.3 Value cocreation

Value cocreation has become a relevant area of business, referring to the phenomena of individualization and singularization as shown above. Academically value cocreation can be anchored in the theory of service dominant logic (SDL) (Vargo , Lusch, Akaka, & He, 2010; also see www.sdlogic.net). The SDL starts with the premise, that the traditional differentiation between goods and services is no longer valid. Instead, the theory considers the exchange processes between market participants, for example, suppliers, manufacturers, and customers. Physical goods are seen as tangible distributors of the service, or in other words, the physical goods are "service suppliers." The concept takes into account the fact that consumers do not demand physical goods, but rather the service associated with their consumption, for example, healthy eating and enjoyment while eating a fish (Drengner, 2020). Getz, Robinson, Andersson, and Vujicic (2014, p. 173f) give a profound overview of the SDL in the context of food tourism, whereas examples for value cocreation are

- Cooking classes (e.g., Kochbox, n.d.).
- Kitchen parties, where visitors look over the chefs' shoulders in their natural surroundings and consume the prepared foods standing directly in the kitchen or presented as a flying buffet menu (e.g., Deidesheimer Hof, n.d.).
- Themed stagings in which the food matches, for example, the hosts biography (e.g., The Fat Duck, n.d.).
- Outstanding food experiences such as dining in underwater restaurants, where consumers eat sitting in the middle of the scenery of the underwater world; here the location is the vehicle for an extraordinary luxury food experience (e.g., Under, n.d.).

In summary Fig. 16.2 shows the discussed drivers of change that will sharp food experiences in a postmodernists world.

16.2.4 Food experience in a post Corona future

In March 2020, the Corona virus Covid-19 led to economic and social shutdown worldwide, which resulted in severe social isolation of people. Travel and business

Figure 16.2 Conceptual framework of the drivers of change.

trips were prohibited and cancelled worldwide, professional and private life was largely transferred to home office. It is assumed that the social isolation and suddenly enforced digitalisation in all life areas triggered by Covid-19 will also have an influence on eating culture. While a "destructuring" of food consumption was experienced in western societies in the pre-Corona era (Zukunftsinstitut, 2020), the corona-induced dissolution of everyday life triggered a renaissance and a return to traditional eating habits. Regular eating with people provides structure and stability to everyday life (Albrecht, Kara and Lobenstein, 2019), especially in uncertain situations. Thus, due to the Corona regulations, social distancing and the closure of restaurants it was observed, that people mainly ate at home and regularly sat down at a table again. In a country like Germany, in which take-away was allowed during the restaurant closure, people were forced to return to their home to consume the food. Others did not want to get a delivery or take-away and re-discovered cooking and baking as a leisure or family activity. It is to be assumed that eating together, both in commercial and non commercial spaces at home and away from home, will experience a new appreciation as "Quality Time" in the post-Corona period. Especially the developments of the virtual common experience mentioned in this chapter gained considerably in importance during the long period of social isolation. The discussed trend of snackification and the destructuring of everyday eating (Zukunftsinstitut, 2020) could, on the other hand, lose significance and lead to a more structured common eating experience. Delivery services and so-called ghost restaurants have clearly gained in prominence during the Corona crisis. Also in view of globalisation, the risks and limits of which became clear in the course of the Corona crisis on many levels, such as along global supply chains and in tourism, a localisation of consumption can be observed, which is expressed in the strengthening of local identities and autonomies (Anthes *et al.*, 2020). The Corona crisis can significantly intensify this development of re-regionalisation, strengthening regional networks and local food production. Weeks of social distancing transformed the private living space into a digital media laboratory (Reitz, 2020), in which the public and the professional world penetrated through digital channels into the last corners of the private living space. The world-including the professional world-moved into our homes digitally and at the same moment the public space was empty. Social interactions of all kinds were banished to the private sphere of the home and, above all, to digital channels. This shift results in an ironical distortion of the private and public. Consequently, Corona has also shown in a precarious way that once private space has become public space through digitalization and that public space, e.g. a restaurant, can become a new place of retreat in which the digital is deliberately excluded while celebrating a meal together.The future of shared dining experiences in a post-Corona era remains a large field of research, which has only been addressed here with regard to digitalization and localization as the driving forces of change. It is likely that eating experiences will in future be more strongly influenced by the longing for security, community and structure. It can be assumed that food is staged as an anchor of security. Experiences in a sophisticated, quiet, familiar atmosphere, both at home and in restaurants, is likely to grow in importance.

16.2.5 Thinking out of the box

As mentioned above, future is not the optimized present. It is therefore advisable that researchers exploring the future of the food experience detach themselves from established industry thinking and let themselves be surprised and inspired by other disciplines and structures, for example, by lateral entrants. Art and food design, for example, can make a relevant contribution, that leads to irritation and thus opens up spaces for new narratives and solutions (Zukunftsinstitut, 2020, p. 42). For example, the Dutch choreographer Nanine Linning worked in 2018 at the Heidelberg Theater together with the food designer Remco Vellinga in her choreography "Khora." Visitors of the dance production received three different small samples, specially designed for the sensory perceptions of taste, haptic tingling and crackling as an acoustic stimulus, inspired by the elements water, fire, and air. The aim of this food experience was to intensify the sensory experiences during the dance companies' performance and led to a hypersensory experience. At the end of the play, the company surprised with a special performance: the dancers hung upside down from the ceiling of the stage, serving specially created "Khora Sparkling Wine" for the visitors and thus addressing the element of earth by reversing gravity (Theater

Heidelberg Germany, 2018, own visit on April 26, 2018). In this context explorative questions arise, such as:

- Can a sound create a taste? How does the favorite jazz or classical music playlist taste then?
- Could a data mix of, for example, sound, light, movement, and text create food as known from 3D-printing?
- Can we heal ourselves with food, whether we can eat our way out of situations and into new situations? Why not get recommendations for certain moods, individual food ingredients that take the eater out of his world of thoughts and moods, for example via app from preferred retailers or restaurants?
- And how would that taste individually?

Combined with the abovementioned trend of urban food, vertical gardening concepts could bring food as an artwork to the walls of our homes and workplaces (e.g., "the edible living room" see ponix-systems.at, as cited by Zukunftsinstitut, 2020, p. 85).

Even the dancers hanging upside down and serving champagne in the theater in Heidelberg introduce a radical change of perspective and override learned patterns of action which lead to the following questions:

- What does it mean to be served?
- What special skills does the person in charge of service have?
- Is service a matter of course or an expression of an exclusive, personal relationship between guest and host?
- Under what conditions must the service people work?
- How can a consumer and guest take responsibility for the working conditions?
- Where are the boundaries between service and consumption? Are there any boundaries at all? Or is every guest also part of the whole team?
- Can the guest cook by himself in a restaurant?
- Can the restaurant kitchen be just a location in which the guest cocreates?
- Can the guest bring home-grown ingredients to the restaurant and the cook creates the experience?
- Can retailers or restaurants offer self-cooking stations?

This staged food experience could also lead to the idea, how food experiences might develop in future: interpreting the service of offering food and the table as a stage along the ideas of Pine and Gilmore (1990) and value cocreation (Vargo & Lusch, 2017) the physical boundaries between the supplier and guest will be removed.

Furthermore following the above discussed development further, one comes across with the idea of a "Super App," that horizontally integrates a variety of services and completely maps the everyday life of consumers in one single application. From messengers, online shopping and on-site payment systems to services such as mobility services, overnight accommodation as well as food orders in alignment with health and movement data, preferences, and dietary requirements. Personal digital profiles based on user behavior and mood analysis from voice recordings, which are continuously and colearningly created, could also provide recommendations for food. A personal assistant, comparable to, for example, Siri and Alexa, digitally coordinates the harvest of fresh foods and purchase of other ingredients, the delivery, and the activation of cooking machines in smart homes, table seat reservations in the desired ambience and the desired food with people of suitable mood and interests, or people who live nearby and are invited home to the kitchen. This scenario is motivated by the idea of process automation based on smart data: from the demand to individual menu or restaurant suggestions, the order or the booking, the food supply, the preparation and invitation of neighbors, family and friends, nearby or via holograms for example. The question which remains about the likelihood that this scenario will occur is whether consumers will approve this idea and accept a machine to coordinate their lives for their own benefit. Moreover, data protection issues would have to be brought in line with such scenarios.

16.2.6 Managerial implications

The future of food experiences remains a huge field of research, which has only been touched on here with regard to some drivers of change. However, these drivers of change can give practitioners ideas to create innovative concepts for food experiences in the field of digital versus physical experiences, slow versus fast food, quality time meals versus on-the-go and take-away—both healthy and highly individualized—single versus family and friends style, and hypersensory versus simple food and taste experiences.

References

Aaltojärvi, I., Kontukoski, M., & Hopia, A. (2018). Framing the local food experience: A case study of a Finnish pop-up restaurant. *British Food Journal*, *120*(1), 133–145.

Albrecht, H., Kara, S., & Lobenstein, C. (2019). Zu Tisch. *Die Zeit*, 27–29.

Ali, N., Ferdinand, N., & Chidzey, M. (2017). In N. Ferdinand, & P. J. Kitchin (Eds.), *Event design* (pp. 67–93). London: Events management – An international approach.

Amazon. (n.d.). *Amazon physical store locations*. <https://www.amazon.com/find-your-store/b/?node = 17608448011> Accessed 09.03.20.

Anthes, D., Carsten, S., Dettling, D., Gatterer, H., Horx, M., Horx-Strathern, O., & Tewes, S. (2020). *Die Welt nach Corona*. Edited by H. Gatterer and M. Horx. Henrich Druck und Medien.

Bell, C. (1997). *Rituals: Perspectives and dimensions*. New York: Oxford University Press.

Beller, S. (2019). Den Berg kochen. Was in den Bergen wächst, bleibt in den Bergen. , *Südtirol entdecken, erleben, genießen.*, 20–27.

Björk, P., & Kauppinen-Räisänen, H. (2016). Local food: A source for destination attraction. *International Journal of Contemporary Hospitality Management*, *28*(1), 177–194.

Brand, R., & Rocchi, S. (2011). *Rethinking value in a changing landscape. A model for strategic reflection and business transformation. A philips design paper*. <www.researchgate.net/publication/272941478_Rethinking_value_in_a_changing_landscape_A_model_for_strategic_reflection_and_business_transformation>. Accessed 30.06.20.

Casini, L., Contini, C., Romano, C., & Scozzafava, G. (2015). Trends in food consumptions: What is happening to generation x? *British Food Journal*, *117*(2), 705–718.

Chen, Q., & Huang, R. (2018). Local food in China: A viable destination attraction. *British Food Journal*, *120*, 1.

Chitakunye, P., & Takhar, A. (2014). Consuming family quality time: The role of technological devices at mealtimes. *British Food Journal*, *116*(7), 1162–1179.

Coca Cola. (n.d.(a)). *Coca Cola freestyle*. <https://www.coca-cola-deutschland.de/uber-uns/karriere/wir-sind-coca-cola/coca-cola-freestyle-cr> Accessed 09.03.20.

Coca Cola. (n.d.(b)). *Coca Cola freestyle demo*. <https://www.youtube.com/watch?v = U18uJj0GNco> Accessed 09.03.20.

Deidesheimer Hof. (n.d.). *Küchenparty mit Kopfkissen*. <https://www.deidesheimerhof.de/de/home/kuechenparty-mit-kopfkissen> Accessed 09.03.20.

Drengner, J. (2020). Service-dominant logic – Konzept und Implikationen für das Marketingmanagement. *Business + Innovation*, *3*(4), 8–15.

Ertugral, S. M., Cetin, G., Karagoz, A., Balık, M., Dincer, F. I., & Sengel, T. (2015). Tourists approach to local food. *Procedia – Social and Behavioral Sciences*, *195*, 429–437.

Getz, D., Robinson, R., Andersson, T., & Vujicic, S. (2014). *Foodies and food tourism*. Oxford: Goodfellow Publishers Ltd.

Hay, B. (2015). The future of dining alone: 700 friends and I dine alone. In I. Yeoman, et al. (Eds.), *The future of food tourism. Foodies, experiences, excklusivity, visions and political capital* (pp. 194–207). New York: Channel View Publication.

Hello Fresh. (n.d.). *Tschüss Alltagsstress. Hello fresh*. <https://www.hellofresh.de/> Accessed 13.03.20.

Kauppinen-Räisänen, H., Gummerus, J., & Lehtola, K. (2013). Remembered eating experiences described by the self, place, food, context and time. *British Food Journal*, *115*(5), 666–685.

Kim, J. H., & Jang, S. C. (Shawn) (2016). Determinants of authentic experiences: An extended Gilmore and Pine model for ethnic restaurants. *International Journal of Contemporary Hospitality Management*, *28*(10), 2247–2266.

Kim, S., & Iwashita, C. (2016). Cooking identity and food tourism: The case of Japanese udon noodles. *Tourism Recreation Research*, *41*(1), 89–100.

Kochbox. (n.d.). *Kochbox – Cullinary rebellion*. <https://kochbox.com/> Accessed 09.03.20.

Lehner, M. (2015). Retail store influence on sustainable consumption behaviour. *International Journal of Quality and Service Sciences*, *7*(4), 404–423.

Lin, L., & Mao, P. C. (2015). Food for memories and culture – A content analysis study of food specialties and souvenirs. *Journal of Hospitality and Tourism Management*, *22*, 19–29.

Mason, M. C., & Paggiaro, A. (2012). Investigating the role of festivalscape in culinary tourism: The case of food and wine events. *Tourism Management*, *33*(6), 1329–1336.

Mikrolandwirtschaft.org. (n.d.). *Mikro Landwirtschaft – gemeinschaftlich Gemüse anbauen – direkt in der Nachbarschaft*. <https://mikrolandwirtschaft.org/infos/about> Accessed 09.03.20.

Mirosa, M., & Lawson, R. (2012). Revealing the lifestyles of local food consumers. *British Food Journal*, *114*(6), 816–825.

Moscardo, G., Minihan, C., & OLeary, J. (2015). Dimensions of the food tourism experience: Building future scenarios. In I. Yeoman, et al. (Eds.), The future of food tourism:

Foodies, experiences, exclusivity, visions and political capital. (Aspects of Tourism) (pp. 208–222). Bristol: Channel View Publications.

Munjal, S., Sharma, S., & Menon, P. (2016). Moving towards "Slow Food", the new frontier of culinary innovation in India: The Vedatya experience. *Worldwide Hospitality and Tourism Themes*, *8*(4), 444–460.

Pandey, M., & Verma, R. (2015). Factors influencing the buying behavior of consumers towards organized retail stores in Jalandhar, Punjab. *Journal of Marketing Management*, *14*(1), 51–65.

Parsa, H. G., Self, J. T., Njite, D., & King, T. (2005). Why restaurants fail. *Cornell Hotel and Restaurant Administration Quarterly*, *3*, 304–322.

Pine, B. J., & Gilmore, J. H. (1990). *The experience economy – Work is theatre & every business a stage*. Boston, MA: Harvard Business School Press Boston.

Privitera, D., Nedelcu, A., & Nicula, V. (2018). Gastronomic and food tourism as an economic local resource: Case studies from Romania and Italy. *Geojournal of Tourism and Geosites*, *21*, 1.

Reitz, E. (2020). Warum wir das Kino brauchen. *Die Zeit*, *24*, 50.

Rewe. (n.d.). *Rewe to go*. <https://togo.rewe.de/> Accessed 09.03.20.

Rinaldi, C. (2017). Food and gastronomy for sustainable place development: A multidisciplinary analysis of different theoretical approaches. *Sustainability (Switzerland)*, *38*(2), 1–25.

Schwert, C. (2018). Strohfeuer oder longseller. *Gastronomie & Hottelerie*, *11*, 16–19.

Scott, M. L., & Vallen, B. (2019). Expanding the lens of food well-being: An examination of contemporary marketing, policy, and practice with an eye on the future. *Journal of Public Policy and Marketing*, *38*(2), 127–135.

Seyler, N. J. (2020). *Megatrends and sustainability* (pp. 5–23). Wiesbaden: Springer Gabler.

Sidali, K. L., Kastenholz, E., & Bianchi, R. (2015). Food tourism, niche markets and products in rural tourism: Combining the intimacy model and the experience economy as a rural development strategy. *Journal of Sustainable Tourism*, *23*(8–9), 1179–1197.

Sims, R. (2009). Food, place and authenticity: Local food and the sustainable tourism experience. *Journal of Sustainable Tourism*.

Statista. (2020). *Anzahl der Einpersonenhaushalte in Deutschland von 1991 bis 2018*. <https://de.statista.com/statistik/daten/studie/156951/umfrage/anzahl-der-einpersonen-haushalte-in-deutschland-seit-1991/>Accessed 30.06.2020.

Steffen, A., & Doppler, S. (2019). Building consumer trust and satisfaction through sustainable business practices with organic supermarkets. In J. Byrom, & D. Medway (Eds.), *Case studies in food retailing and distribution*. Duxford: Elsevier Woodhead Publishing.

The Fat Duck. (n.d.). *The trip: Your Itinerary*. <https://www.thefatduck.co.uk/the-itinerary> Accessed 12.03.20.

The Worlds 50 Best Restaurants. (2017). Heston Blumenthal: "You need the freedom to fail". <www.theworlds50best.com/blog/News/heston-blumenthal-fat-duck-you-need-the-freedom-to-fail.html> Accessed 10.05.18.

Tsai, C. T. S. (2016). Memorable tourist experiences and place attachment when consuming local food. *International Journal of Tourism Research*, *18*, 6.

Under. (n.d). *Under – A restaurant five and a half meters below the surface in Norway, exposing you to the wonders beneath the sea*. <https://under.no/> Accessed 09.03.20.

Vargo, S. L., & Lusch, R. F. (2017). Service-dominant logic 2025. *International Journal of Research in Marketing*, *34*(1), 46–67.

Vargo, S., Lusch, R. F., Akaka, M. A., & He, Y. (2010). Review and assessment of service dominant logic. *Review of Marketing Research*, *6*, 125–167.

Yeoman, I., McMahon-Beattie, U., & Wheatley, C. (2015). The future of food tourism: A cognitive map(s) perspective. In I. Yeoman, et al. (Eds.), *The future of food tourism: Foodies, experiences, exclusivity, visions and political capital* (pp. 237–278). Bristol: Channel View Publication.

Yusuf, S. (1999). The changing development landscape. *Finance and Development*, *36*(4), 15–18.

Zhang, J., Liu, H., Sayogo, D. S., Picazo-Vela, S., & Luna-Reyes, L. (2016). Strengthening institutional-based trust for sustainable consumption: Lessons for smart disclosure. *Government Information Quarterly*, *33*, 552–561, Wiesbaden: Elsevier Ltd.

Zukunftsinstitut. (2018). *Megatrend-Dokumentation*. Frankfurt am Main: Author.

Zukunftsinstitut. (2020) *Hanni Rützlers food report*. Frankfurt: Author.

Zukunftsinstitut. (n.d.). *Dossier food*. <www.zukunftsinstitut.de/dossier/dossier-food/> Accessed 17.02.20.

Index

Note: Page numbers followed by "*f*" and "*t*" refer to figures and tables, respectively.

9780128177921